DEFENCE
IN UNCL

CRANFIELD DEFENCE MANAGEMENT SERIES
(*Cranfield University Department of Defence Management and Security Analysis*)
Series Editors: Trevor Taylor and Teri McConville
ISSN 1740-3073

1. *Human Resources Management in the
British Armed Forces* (2001)
edited by Alex Alexandrou, Richard Bartle and Richard Holmes

2. *New People Strategies for the British Armed Forces* (2002)
edited by Alex Alexandrou, Richard Bartle and Richard Holmes

OTHER RELATED TITLES

*The British Army, Manpower and Society into the
Twenty-First Century* (2000)
edited by Hew Strachan

*Democratic Societies and their Armed Forces:
Israel in Comparative Context* (2000)
edited by Stuart A. Cohen

DEFENCE MANAGEMENT IN UNCERTAIN TIMES

Editors

Teri McConville
and
Richard Holmes

FRANK CASS
LONDON • PORTLAND, OR

First published in 2003 in Great Britain by
FRANK CASS PUBLISHERS
Crown House, 47 Chase Side, Southgate
London, N14 5BP, England

and in the United States of America by
FRANK CASS PUBLISHERS
c/o ISBS, 920 NE 58th Avenue, Suite 300
Portland, Oregon, 97213-3786

Website: www.frankcass.com

British Library Cataloguing in Publication Data

Defence management in uncertain times: investing in the future
– (Cranfield defence management; no. 3)
1. National security – Great Britain 2. Great Britain –
Military policy 3. Great Britain – Armed Forces –
Recruiting, enlistment, etc.
I. McConville, Teri II. Holmes, Richard, 1946–
355'.033041

ISBN 0-7146-5522-8 (cloth)
ISBN 0-7146-8414-7 (paper)

Library of Congress Cataloging-in-Publication Data

Defence management in uncertain times / editors, Teri McConville
and Richard Holmes. – 1st ed.
    p. cm.
Includes bibliographical references and index.
  ISBN 0-7146-5522-8 – ISBN 0-7146-8414-7
  1. Great Britain – Armed Forces – Management. 2. Military art
and science – Great Britain. 3. World politics – 21st century.
I. McConville, Teri, 1954– II. Holmes, Richard, 1946– III. Title.
  UB58 21st century .D44 2003
  355.3'0941'0905 – dc21
                                                          2003005220

Typeset in 10.5/12pt ZapfCalligraphica by Frank Cass Publishers
Printed in Great Britain by MPG Books Ltd, Bodmin, Cornwall

Contents

The Authors

RICHARD BARTLE was an Army Officer for 23 years. He retired in the rank of Lieutenant Colonel and is now a Cranfield University lecturer in the Department of Defence Management and Security Analysis at the Royal Military College of Science. He has published widely on Organisational Behaviour and Human Resource Management in the military sphere and is the co-editor of two recent books on HRM in the Armed Forces.

ANDY BOLT is an Engineer Officer in the RAF and is currently studying for a Masters in Defence Administration. He has enjoyed tours managing communications and information systems on a flying station, controlling satellites, teaching engineers, and most recently, coordinating overseas deployments. He maintains a good work-life balance by mountaineering, playing sport and spending time with his wife, Helen.

BRIAN HOWIESON joined the Royal Air Force in 1987. He has completed operational tours on both the Nimrod MR2 and the Nimrod R aircraft and was involved in the 1991 Gulf War; the Kosovo War and air operations against Northern and Southern Iraq. In addition, he has completed a tour as a flying instructor at the Royal Air Force College Cranwell. Brian held a Defence Fellowship at Heriot-Watt University in Edinburgh and has worked in the Joint Defence and Concept Centre at the UK Defence Academy. He is currently on secondment to the National Mentoring Consortium in the East End of London. He is currently working on his PhD at the University of Edinburgh and his research interests include the influence of leadership on goal-setting and strategy.

HOWARD KAHN obtained his MA (Political Economy and Modern History) from the University of Glasgow and his MSc (Occupational Psychology) and PhD (Occupational Stress) from the University of Manchester Institute of Science and Technology. After a career as a computer and business analyst with British Steel and Lloyd's of London (Underwriters), he lectured at Manchester Polytechnic in Systems Analysis. He is currently Senior Lecturer in Organisational Behaviour in the School of Management and Languages at Heriot-Watt University, Edinburgh. Howard has written more than 50 books, book chapters, and academic papers, and regularly contributes to the media on aspects of people at work. He is also cited as an expert witness in stress-at-work litigation. His current interests include the development of leadership in the public and private sectors.

CHARLES KIRKE is a serving officer in the British Army. He joined the Royal Artillery in 1970 and has served at regimental duty with field artillery and in recruit training, and on the staff in MoD (Operational Requirements), HQ UK Land Forces (Artillery Branch), and RMCS (Directing Staff), and is currently the military member of the Defence Science and Technical Laboratory (Dstl) Human Sciences Team. He took a degree in Social Anthropology in 1974, carried out a Defence Fellowship at Cambridge University in 1993/4, and is a PhD Candidate with Cranfield University (RMCS) researching social structures in the regular combat arms units of the British Army.

TERI McCONVILLE gained practical management experience in the public sector. Following professional training and a commission in Princess Mary's Royal Air Force Nursing Service, she moved into the NHS where she worked as a nursing officer and Sister Tutor. Teri gained a PhD from the University of Plymouth before joining the Defence Management Group at the Royal Military College of Science to teach Organisational Behaviour. She is currently involved in research activity surrounding the debate over women in combat, and the 'piggy in the middle' effect of line management.

PATRICK MILEHAM is Reader in Corporate Management and Staff Governor, University of Paisley. He served in the Regular Army from 1964 to 1992 and holds degrees from the Universities of Cambridge and Lancaster. As Associate Fellow of both the Royal Institute of International affairs and Royal United Services Institute he has been instrumental in organising conferences on operational military ethics and has conducted much research into military motivation, leadership and morale in the UK and overseas. Among his many publications on historical and current topics are 'Will They Fight and Will They Die?' in *International Affairs* (2001) and 'Morale in the Armed Forces' in *RUSI Journal* (2001).

GEORGINA NATZIO formerly Deputy Editor of the official journal of the veterinary profession, *The Veterinary Record*, entered publishing aged 18. Her first introduction to military, rather than veterinary science, came via animal husbandry in a pigsty belonging to 58 Medium Regiment RA, aged 6. Later returning to informal study of military science after retirement from publishing aged 28, her work was generously guided and tutored by three World War II veterans from the Royal Navy and the Army. Her first review essay in this context, 'The Future of Women in the Armed Forces', *RUSI Journal*, December 1978 was followed by 'British Army Servicemen and Women 1939–45, their Selection, Care and Management', *RUSI Journal*, February 1993, and 'On the Military Significance of Being Female', *British Army Review* No. 126, Winter 2000–2001. She has written and reviewed for both journals since becoming a member of the RUSI in 1974.

JACK SPENCE is Academic Adviser to the Royal College of Defence Studies, London and Visiting Professor in the Department of War Studies, King's College, London. He was also the former director of studies at the Royal Institute of International Affairs, London. He is grateful to Captain Ian Richardson, a Royal Navy member of the Royal College of Defence Studies (2000/01), for some helpful comments on the first draft of this chapter.

TREVOR TAYLOR is the Head of the Department of Defence Management and Security Analysis at Cranfield University's faculty at the Royal Military College of Science in the UK. He was previously Professor of International Relations at Staffordshire University and between 1990 and 1993 was Head of the International Security Programme at the Royal Institute of International Affairs in London. He is also a past Chairman of the British International Studies Association and has been Visiting Professor at the National Defence Academy in Tokyo. He was educated at the London School of Economics (BSc (Econ) and PhD) and Lehigh University (MA) in Pennsylvania.

JAMES YORK was commissioned into the Royal Anglian Regiment in 1991. He has served as a platoon commander in armoured, mechanised and airmobile infantry roles. Regimental duty has also included a tour as a reconnaissance platoon commander and as a company commander with the 1st Battalion Royal Anglian Regiment in Londonderry. Outside of the Regiment, Major York served in a training regiment and as SO3 Media Operations at PJHQ. He has recently completed an MA at the Royal Military College of Science and attended the Joint Services Command and Staff Course prior to taking up his current appointment of SO2 Reserves within the Directorate of Army Staff Duties. James is married to Mandy and they have a young son Tom.

Introduction

May you live in interesting times
(Ancient Chinese curse)

In defence, as in any other field, management occurs at all levels of an organisation and governs the full range of organisational activity. It begins with the formulation of policy at the highest levels of command and government, and extends to the direction and control, sometimes in fine detail, of all aspects of Service life. Managers need to interpret the environment in order to plan, organise, direct, coordinate and control the efforts of their organisations.

However, being a very public service, defence and its management is always open to scrutiny from government, international bodies such as the United Nations, and from wider populations via the mass media. Hence, defence management is carried out under the gaze of the population who can be swift to criticise, even when they do not fully understand what is expected of defence services.

For most of their history, armies, navies and air forces had a fairly simple and practical task – to control the means of legitimate violence and to fight on behalf of the state. In recent years that task has broadened, reflecting the enormous political, social, technological and economic changes that have occurred in the world. For many commentators the fall of the Berlin Wall in 1989 exemplified the extent of world change. The collapse of state communism and the end of the Cold War marked the movement of defence from the modern into a post-modern era.

One year following 11 September 2001 (the time of writing) it seems that that date may become a more definitive moment for it has truly changed the world. Unlikely alliances have been formed to combat international terrorism and, as the 'enemy' is unidentified and indefinable, the very nature of war has been transformed, possibly forever.

The demands upon defence services are potentially infinite, but principals will always need to manage within the constraint of finite resources. Among other things, there is movement toward 'professional' militaries as many nations seek an end to conscription. Jointery, the merging of support functions into multi-service 'purple' organisations, seems to have become a necessity, rather than a simple idea. Outsourcing of many key functions is becoming commonplace.

The result is that the Armed Forces, which were once self-servicing institutions are now dependent upon numerous others to support their functions, which many see as disadvantageous to the Services. On the other hand, professional military officers can be freed from the minutiae of organisational management to concentrate on their core functions.

The changes that are occurring are not confined to war fighting[1] nor even to the Armed Services and their roles. They are part of a wider, but relentless, movement concerning informatics, globalisation, and public sector reform. The post-modern world is typified by a breaking down of accepted norms, a relaxation of established structures and greater regard for individuality. For disciplined forces, who must necessarily act in concert with others, and who serve a collective ideal, the world is becoming a strange place where uncertainty is a fact of life.

Moskos *et al.*[2] note that a post-modern military exhibits five defining features. There is a blurring of the demarcation between civilian and military institutions; and, within the services, there is less distinction between combat and support roles. The armed forces may be called to serve in countries other than their own and frequently as part of multinational operations. Their changing and changeable function now encompasses many forms of operation other than war. Whether or not we choose to accept the post-modern label, such shifts in structure and purpose are apparent in myriad nations and cultures and are, of themselves, sources of uncertainty.

Defence managers, like their counterparts in both the public and private sector need to learn to cope with change and the resulting uncertainty. This is no easy task for uncertain situations mean that there are no sure answers or solutions. This volume represents the attempts of its contributors, military and academic, to assist in the process. It should be noted that the views expressed by the various writers are their own and that such views do not represent any official position.

The book opens with a chapter from Jim York who, as a serving infantry officer, is familiar with uncertainty. He recognises that this condition that is not unique to defence but within that arena it is the uncertain environment that offers opportunity for victory. His chapter approaches the subject in three parts.

He begins by demonstrating how dealing with uncertainty has led to the invention of industries such as insurance; and to the development of a variety of responsive organisational structures.

Second, his work moves on to test the language of uncertainty as a way of describing military and, particularly, command activity. Information technology is an important part of twenty-first century life and is now making a firm impression on military operations.

The final part of Major York's chapter offers a view on how such technology should be applied if it is accepted that uncertainty is the main problem of war. More information, he suggests, is not necessarily the answer to reducing uncertainty. Rather, it is for commanders to come to terms with uncertainty. They must be able to work in uncertain environments if they are to exploit the opportunities that it presents.

On a more specific topic, Jack Spence explores the extent to which the events of 11 September 2001 herald a fundamental change in the structure and politics of international society. He analyses the nature of terrorism and, in particular, how far the activities of al-Qaeda are a radical departure from the practice of orthodox terrorism in the post-1945 period. He gives special attention to the impact on interstate relations, movements toward multilateralism, and Britain's role in mobilising world support. The role of public opinion in Western states is also considered, together with consequences for the maintenance of civil liberties. Some attention is given to what, if anything, can be done to close the North–South divide as well as the impact on international institutions such as the United Nations and the European Union.

The chapter concludes with a brief survey of the possible consequences of 11 September for the academic study of international relations. It is noteworthy that this work was originally written within a month of the attacks upon the

USA. The valuable insights generated by Professor Spence are a remarkable example of the value of strategic analysis.

Teri McConville approaches the aftermath of 11 September 2001, and particularly the 'war on terrorism', from an academic viewpoint. Her chapter concerns the decision-making processes that occur in high-profile policy-making groups. This discussion arises from her concerns that there was an underlying pathology in the decision-making process. The particular pathology, groupthink, has been implicated in many catastrophic policies in the past, and defence managers need to be aware of how it can influence decision- and policy-making activities.

Dr McConville uses documentary evidence of the world reaction to the terrorist attacks on the USA, and the resulting offensive against the Taliban, to test whether groupthink had occurred within the White House and other international policy-making bodies. She also discusses strategies for reducing the risk of such pathology in the formulation of policy decisions.

Considering a major policy issue, Trevor Taylor turns our attention to the concept of what the British are calling 'jointery' – the integration of separate force activities. He believes that that concept, and the British experience, will be of increasing interest to other states as they seek to optimise the effectiveness of their armed forces and the defence capabilities within the increasing constraints on defence budgets.

While jointery can cover the preparation, operations and resourcing of military forces (most obviously training) and, closely related, the management of the resources used by the forces,[3] it is a process that might included a range of government and non-government actors. Pressure for jointery comes from the need for efficient and effective military performance but as such pressure increases, it is likely to become a continuing process rather than an achievable end-state.

Professor Taylor likens jointery with the concept of European integration and particularly explores the notion of spillover, where cooperative effort in one area widens into related areas spreading the benefits of the original activity. However, he emphasises, jointery can only be successful within the framework of a coherent and specific policy that provides direction to those shaping armed forces.

Organisations in the public sphere rely on the efforts of a motivated and skilled workforce to achieve their primary functions. The way that people are recruited and trained is therefore of vital importance, especially as demographic trends erode the pool of potential recruits. A sizeable portion of this volume is concerned with human resource management (HRM) issues.

Since the end of the Cold War, nation-states have increasingly moved away from conscription, favouring the concept of a professional military. Seldom does anyone question the basic notion of 'professional' in respect of the Armed Forces although the concept contains a combination of conceptual, personal and moral components.[4] Noting that professions only exist with the consent of the wider population, Patrick Mileham offers a thoughtful chapter to catalogue and explore the underlying concepts and practices associated with the often abstract idea of professional military.

Dr Mileham's aim is to explore, rather than define, professionalism. He presents and discusses a brief catalogue of assumptions that are generally included in popular notions of the concept. He then explores their implications for 'the profession of arms'.[5] and particularly within established or emerging democracies. He notes that trends to post-modernity, by eroding traditional values, could undermine the implicit qualities of professionalism but contends that certain truths will endure.

Howieson and Kahn are concerned with how changes in the wider environment have affected civil-military relations and recruitment. They note that while the Services are working hard to modernise their professional practices, they have not been addressing the cultural and moral changes that are occurring in the wider society. Conspicuous among those is a declining regard for authority and social institutions; a movement toward short-term careers, a declining role for the state, and the so-called information revolution.

Such macro-environmental changes affect the whole population and, particularly the young people that are potential recruits for the Services. In their chapter they report on their large-scale research into the attitudes of graduates towards careers in the Armed Forces. This report indicates that most of their sample did not view the military as representative of society and that the public perception of the Armed Forces as employers is unfavourable.

Clearly, such perceptions and attitudes have implications for recruitment and retention of personnel and these authors make several recommendations for those concerned with such matters.

For those who decide on the military for their future career, there is a culture shock waiting. This is nothing new; indeed, much of basic training is about socialising young people into their new environment and culture. What is new is that such recruits are joining from a social culture that is notably different from that of former generations. Charles Kirke notes that contemporary youth culture displays the features of post-modernism being unstructured and egocentric. He contrasts that with the structured, disciplined, environment presented by the military.

The resulting culture gap, which must be crossed by latter-day recruits, is greater than anything experienced by their forebears. The only help they have is from recruiters and trainers who are already socialised into this alien environment. Lt. Colonel Kirke questions whether the resulting difficulties could contribute to some recruits' inability to complete their basic training. Perhaps some future work is indicated to develop some form of 'cultural bridge' to help recruits to cross the gap during the early stages of their training.

Once recruits have been attracted to the armed forces as a career, (HRM) policies need to address the problem of retention. Andy Bolt addresses HRM issues from this perspective, by considering the nature of work-life balance. He points out that in organisations where membership is institutional, the benefits on offer must address both ends of the equation. He develops a model of life satisfaction to plot graphically four dimensions of life interest and to show how these might change over the course of a Service career.

Recruitment can be improved by providing benefits tailored to the needs of the target demographic, namely young people, and communicating that the benefits offered are not confined to the single dimension of monetary reward. Life-satisfaction or work-life balance policies may be instrumental in attracting the current generation of prospective recruits who are less willing to accept the traditional demands of service life by illustrating the advantages that accompany the traditional restrictions.

Retention could also be ameliorated by tailoring the benefits of continuing service to the needs and requirements of married couples and families, that is by fitting the rewards offered to the needs of those that are leaving. As a result of his work Squadron Leader Bolt makes a strong case that the armed forces should seek a competitive advantage in the labour market from its public service ethos and the benefits of the Service life.

Continuing the theme of recruitment and retention, Richard Bartle, addresses the issue of career management. If, as suggested in the Future Career Management of Officers,[6] a new career structure is to be introduced then there are implications for the recruitment and training of Officers. The proposal that officers should be able to specialise into certain areas (Defence Policy, Combat, Logistics, Human Resources and Technical) is appropriate and laudable. It is also bound to bring new problems. One of these, Bartle suggests, is that different personality types are attracted to different careers.

Comparing studies carried out in the USA with his own work at the Royal Military College of Science, Bartle has shown that there is a distinct lack of those types who are said to prefer careers fields such as health care, education and communication. For whatever reason, these personality types are either not attracted to, or are not continuing in, a Service career. Whether this situation may change depends upon the recruiting, promotion and training strategies that are introduced to underpin the new plans for future career management.

Addressing the issue of under-recruitment of service personnel a debate has arisen over whether enlisting or commissioning more women might alleviate the problem. In Britain, such deliberations have recently culminated in a policy decision[7] that women should not be employed in combat arms. Georgina Natzio joins the debate by noting that in both ethical and practical terms, the whole arena of women's employment has been in a state of flux since World War II.

The process of government involves the fusion of national interests and ethical principles into what is often an uncomfortable union. In her impartial, historical review, Ms Natzio notes that the results are often paradoxes, where principles and policies run counter to knowledge and experience. She also notes that under certain circumstances ethical precepts may be altered, or even suspended, to meet the demands of national interests. Perhaps this debate, of all the issues addressed in this book, is most symbolic of the uncertainty that faces the world, and managers of defence in 2003.

Whatever the future of defence, be it warfighting, peace-keeping or some other function, the one thing that is certain is that the environment will continue to bring changes and new challenges. The purpose of defence management is to establish policies, procedures and practices that will allow the armed forces to meet those challenges. So while life may indeed be uncertain, it is bound to be interesting. But then the Chinese had a curse …!

RICHARD HOLMES
TERI McCONVILLE

REFERENCES

1. Kellner, D. *The Politics and Costs of Postmodern War in the Age of Bush II* (University College of Los Angeles www.gseis.ucla.edu/faculty/kelllner/papers/POMOwae.htm, 2002).
2. Moskos, C. C., Williams, J. A. and Segal, D. R. 'Armed Forces after the Cold War' In idem (eds.) *The Postmodern Military: Armed Forces after the Cold War* (Oxford: Oxford University Press 2000).
3. Dorman, A., Smith, M. L and Uttley, A. 'Jointery and Combined Operations in an Expeditionary Era: Defining the Issues', *Defense Analysis*, Vol.14, No.2 (1988) pp.1–5.
4. *British Defence Doctrine*, Ministry of Defence: Joint Warfare Publication, JWP 0-01, 1996 pp.3–5 to 3–14.
5. Hackett, J. *The Profession of Arms* (London: Sidgwick & Jackson 1983).
6. APC *The Future Career Management of Officers*. ECAB/P(00)02, 9 March 2000.
7. UK Ministry of Defence, *Women in the Armed Forces* report by the Women in the Armed Forces Steering Group, May 2002, Ministry of Defence (www.mod.gov.uk).

1

The Quest for Certainty; Coping with Uncertainty

JAMES YORK
Royal Anglian Regiment

'**certain**, adj. Sure: not to be doubted: resolved: fixed: determinate: regular: inevitable'.[1]

One of the strongest themes of writings on the Revolution in Military Affairs (RMA) is the importance of information and the ability to exploit information. Terms such as 'information dominance', 'complete situational awareness' and 'recognised ground picture' are frequently used to describe capabilities that will be delivered by new technology. These terms all express a requirement for a level of certainty that is not new. Military commanders through the ages have sought information in order to give them certainty as to the outcome of their decisions.

Certainty, as defined above, would present few problems as an optimal solution would present itself. However, the key is not creating a theoretical information utopia, but rather dealing with the reality of imperfect information and thus having to act in a situation of uncertainty. How individuals and organisations approach dealing with uncertainty drives how they gather information, which is dependent on what they understand information to be. By establishing a framework of definitions and understandings, the actions of historical commanders can be analysed with respect to their approach to certainty. Several authors, including van Creveld[2], suggest that the approach taken by leaders and organisations is fundamental to their success or failure.

The application of the framework in this chapter is not an exhaustive survey of history, rather a method to understand commander's actions in the face of certainty's corollary uncertainty. While it may be thought that this could vindicate supporters of a 'mission command' style approach, it will also show that uncertainty can be overcome by ensuring that your opponent is bent to your will. Lawford and Young commented on Napoleon's style that he was, 'not unduly worried about the intentions of his opponents, as he intended to force them to conform to his own view.'[3] This comment suggests that there is no set solution, rather different methods for dealing with uncertainty.

A trend that is apparent is the gathering of more information to overcome uncertainty. The RMA may provide the technological leverage to overcome the friction imposed by higher information requirements, but this is still open to debate. This debate will be important, as it will shape the doctrine and equipment of the future armed forces. The dangers of wrongly equipping and training an army have been well illustrated throughout history. An unintended outcome of the application of RMA technology may be a centralisation of command and resources in a manner diametrically opposed to current UK doctrine. The chapter will therefore conclude with the impact the desire for certainty may have on the application of RMA technologies.

CERTAINTY AND UNCERTAINTY

Invoking Clausewitz and van Creveld, Schmitt and Klein[4] assert that certainty is the basic problem of war. In this assertion many military thinkers and practitioners would support them. Clausewitz lists uncertainty as one of the factors making up the 'general atmosphere' of war[5].

Wellington acknowledged that he was working from imperfect information when he commented to his riding companion 'Why, I have spent all my life trying to guess what was on the other side of the hill.'[6]

It is this problem of certainty that makes the outcome of military operations so difficult to predict. In order to achieve certainty military commanders are faced with the task of gathering information about themselves, the enemy and the environment. All these pictures will be incomplete to some extent and therefore there will be uncertainty.

It is not, however, a solely military problem, indeed it could be argued that the quest for certainty has provided the human race with its reason for development. Among other benefits the shift to settled agrarian civilizations in the Middle

East was a response to the uncertainties of a hunter-gatherer lifestyle.

Today the problem of resource allocation in modern economies is a reflection of uncertainty, induced by vast and complex interacting systems. Two major schools of thought seem to have arisen in response to this problem.

One view based on a Marxist interpretation sees central resource control and allocation as the solution.

A second, and now arguably predominant view, suggests that in such a complex situation hierarchical central control is not possible. This is largely due to 'the role played by the transmission of information'[7] which will not enable central direction to function efficiently. Unsurprisingly as the underlying problem is the same, military approaches to uncertainty could be equally divided into centralist and distributive solutions.

Humans have sought to quantify and manage uncertainty throughout history. Early civilisations were content to accept fate as a decision-maker. Seamus Heaney describes the seventh century telling of Beowulf's fatal contest with the dragon as a poem 'imbued with a strong intuition of *wyrd* [fate] hovering close, unknowable but certain…'[8].

Bernstein[9] suggested that as human understanding developed over time and by the 1700s a series of writers were contesting the role of fate and seeking to quantify probabilities. From this work sprang industries such as insurance, without which the capital that drove the expansion of the West would not have been staked. Bernstein traces a line of thinking from Pascal in the seventeenth century to twentieth century writers such as Keynes and Knight.

Knight is of particular interest as he drew a clear distinction between risk and uncertainty. Risk was measurable, and therefore could be assigned a probability making it not uncertain. He was, however, critical of man's ability to forecast the future thus, he insisted, '*a priori* reasoning cannot eliminate indeterminateness from the future.'[10] Knight argued that,

> [Any given] 'instance'…. is so entirely unique that there are no others or not a sufficient number to make it possible to tabulate enough like it to form a basis for any inference of value about any real probability in the case

we are interested in. The same obviously applies to most of conduct and not to business decisions alone.[11]

This was written in 1921 in a period when J.F.C. Fuller[12] was to state, 'If we can establish a scientific method of examining war then frequently we shall be able to predict events - future events – from past events.'[13] This disagreement over the predictability of the future has therefore always existed – certainly an indicator that caution should be used when the future is predicted. If we cannot ever predict the future we must be able to cope with uncertainty.

Several writers have demonstrated that organisations can be designed to cope with uncertainty. Much of van Creveld's conclusion to *Command in War* is based on the ideas of Jay Galbraith put forward in *Designing Complex Organisations*. These ideas are based on observing civilian organisations and how they can cope with change and their environment. The main proposition of Galbraith was that, 'the greater the uncertainty of the task, the greater the amount of information that has to be processed between decision-makers during its execution.'[14] High levels of knowledge would allow pre-planning without much interaction between decision-makers.

Conversely, if little is known at the beginning of a task, then as knowledge develops this will lead to a reassessment during the task, which will in turn require reallocation of resources and involvement of decision-makers. This leads Galbraith to conclude that, 'the greater the task uncertainty, the greater the amount of information that must be processed among decision makers during task execution in order to gain a given level of performance'.[15]

Of the three organisational outcomes of this conclusion only one makes sense to the military problem and that is, to 'increase flexibility to adapt to [the] inability to pre-plan'. This may suggest that the solution to tasks involving uncertainty is to be able to process greater amounts of information during the task, something that is promised by today's information technology revolution.

Information processing is key to the problem as information defines uncertainty, 'Uncertainty is defined as the difference between the amount of information required to perform the task and the amount of information already possessed by the organisation.'[16]

4

We seem then to be presented with a simple requirement, identify the information gap, collect information, decide and then act, but this presents two obvious problems.

First, resources cannot be turned on and off like a tap and in large organisations the reallocation of resources can be more problematic than the actual production itself.

Second, there is the idea that information is a transparent, readily available commodity, which it is not. We will first consider the organisational issues raised by resource allocation in uncertain circumstances.

Van Creveld argues, based on Galbraith, that uncertainty is central to the design of a command structure[17]. Given a set performance level, with inadequate information, an organisation has only two choices. The first choice is to increase information flow and the second is to design the organisation so that it can operate with less information.

It is further argued that these two choices lead to one optimal solution and two sub-optimal solutions. By attempting to increase information processing, the centre becomes bloated, unwieldy and defeats itself, through what Van Creveld terms 'information pathology'[18]. Drastic simplification, such as the Greek phalanx, will also lead to a sub-optimal solution in comparison to solutions that subdivide the task and establish forces 'capable of dealing with each of these parts separately on a semi independent basis'.[19]

Based, therefore, on placing uncertainty as central to the military problem, Van Creveld concludes that decisions should be made as low as possible within the organisation, and units required to make decisions must be self-contained. Drawing examples from across history, he also concludes that this rule is independent of developments in technology.

What this analysis lacks is a consideration of the nature of information, even Galbraith's analysis suggests only two sorts of information, known or unknown. This view of information as discrete and finite seems to be an undercurrent in much of the current thinking on the RMA. Schmitt and Klein argue that what we are currently seeing is a data revolution, not an information revolution and the focus has been on increasing information quantity not clarity[20].

Information will increase exponentially as for each new fact gathered, more than one inference may be drawn and these inferences may be drawn together to produce multiple projections.[21] This is not new, what is new is the volume of initial facts that can be gathered depending on the level of resolution the commander wants to go to.

Additionally, some information will be irrelevant, some of it critical. The key is understanding the patterns that present themselves. Directorate Land Warfare (DLW) suggest that the future battlefield may be like a chessboard, with all the pieces located and their capabilities known. The role of the commander is to divine the enemies intentions and act decisively.[22]

This chess analogy is also used by Schmitt and Klein who suggest, 'the thing that separates the chess master from the novice is that the master sees patterns and meaning where the novice sees a disarray of pieces.'[23]

By focusing on the information component of uncertainty, Schmitt and Klein offer several strategies for dealing with uncertainty that share common ideas with van Creveld, but are less prescriptive.[24] In common with van Creveld they identify that more information is not the solution.[25] When faced with uncertainty with a set level of information, a commander can: plan for it; attempt to reduce uncertainty; or attempt to manage uncertainty. Van Creveld would suggest the first two solutions are sub-optimal.

However, in particular circumstances uncertainty can be planned for by utilising reserves and using contingency plans. In order to reduce uncertainty, commanders can identify what information is key and also structure the battlefield to reduce enemy options.

One of Schmitt and Klein's strategies to manage uncertainty shares much with van Creveld thus, 'mission tactics means distributing uncertainty throughout the organisation and thereby decreasing the amount of uncertainty that any given commander must deal with.'[26] They suggest the same division of tasks as van Creveld, but emphasising that sub-division of tasks also distributes uncertainty, allowing it to be managed at an appropriate level.

A further, more aggressive, method of managing uncertainty that echoes Napoleon's style mentioned earlier is also proposed, that of gripping the enemy, thereby reducing his initiative and thus reducing friendly uncertainty.

All these solutions depend on some level of information. We have seen that information is not discrete, but it must be describable if it is to have utility. Darilek suggests that, 'information has two important attributes, *value* and *quality*'.[27] Information is only of value if it adds to the understanding of the situation. The quality of information is dependent on 'accuracy, timeliness and completeness'.

As a theoretical exercise Darilek goes on to develop a probability analysis that allows conditions such as information dominance and superiority to be defined mathematically[28]. Factors considered demonstrate some of the problems with 'information' and include, misreporting, reliability of sensors, terrain occlusion and timeliness. All of these factors will affect the value or quality of the information.

In this model there is also no method to account for enemy deception measures. As the NATO operation in Kosovo showed, simple deception measures can make the quality of information very limited. There is then a discrete set of data we wish to acquire, in order to produce information of a known quality, which will be of value. This would suggest that the art of command is identifying what information will be of value and then gathering it at a set quality level, in order to make decisions.

Although information of value and quality appears to be the bedrock of rational decision-makers, there is a weight of writing that suggests that decision-makers do not utilise information in a rational manner. There is an 'asymmetry' in decision making involving gains and losses. Bernstein quotes Tversky to support this idea, 'It is not so much that people hate uncertainty – but rather, they hate losing.' and he goes on to say 'losses will always loom larger than gains'.[29]

This is possibly why generals such as Patton and Rommel stand out, they are fundamentally different in their outlook. 'Cautious' generals such as Montgomery are too much like ourselves to earn our praise. This asymmetry suggests humans are unable to rationally judge information.

It is also suggested that more information is not as useful as we think and performs the function of a comfort blanket, rather than adding to the value of decisions. Bernstein[30] cites medical research that suggests that as more information is made available, doctors tend towards maintaining their original decision or decide to do nothing – the military implications are obvious and the decision in 1944 to continue with Operation 'Market Garden' could be an example of this tendency in the military.

Schmitt and Klein suggest that decision-makers work on five information cues and no more than ten. This seems perfectly sensible, we acknowledge we cannot have perfect knowledge and in order to allow ourselves thinking space we deliberately limit how much information we will consider. Once again the art is identifying which information to consider.

It is when we become unwilling to accept uncertainty that it becomes a malign influence. Clausewitz recognised that a great commander must be able to accept uncertainty, which along with chance, danger and physical effort made up the atmosphere of war[31]. To Clausewitz it is chance that drives uncertainty:

> War is the province of chance. In no other sphere of human activity is such a margin to be left to this intruder, because no other activity is so much in constant contact with him on all sides. He increases the uncertainty of every circumstance and deranges the course of events.[32]

Importantly, Clausewitz sees this as a continual process and our experience can only be gained by degree over time, 'thus our determinations continue to be assailed by fresh experience'.[33]

This has an implication that is timeless, if you continue to gather information you continue to assail your senses, the faster you gather information the faster you assail your senses. Orienteers only run as fast as they can read a map namely as fast as they assimilate their fresh experience as the terrain around them changes. Here, the advantage that the terrain (excluding the weather) is not dynamic and does not change with the actions of the orienteer as an enemy would.

Another approach to uncertainty is simply to replace it with certainty, at least in our own minds. This can of course lead to the greatest surprises! Kenneth Arrow was very sceptical about man's ability to judge the value of information they held,

> To me our knowledge of the way things work, in society or in nature, comes trailing clouds of vagueness. Vast ills

have followed a belief in certainty...when developing a policy with wide effects for an individual or society, caution is needed because we cannot predict the consequences.[34]

Arrow's views are built partly on his experience as a weather forecaster in World War II. Having realised that the long-range forecasts were no more accurate than a random guess, he suggested dropping them from the workload. In response he was told that, 'The Commanding General is well aware that the forecasts are no good. However, he needs them for planning purposes.'[35]

The general in this example has developed a way of dealing with uncertainty, create certainty and plan to that. If the weather is different from that which is expected then amend the plan. This is not forecasting, it is making an assumption. It is likely that assumptions are shied away from because they involve responsibility. If you forecast good weather for D-Day and there is a storm that is bad luck. If you assume good weather and there is a storm, that is bad planning.

From the foregoing discussion it would appear that uncertainty will always be with us. Much of the modern world is built on this, insurance and the stock markets can only function in an uncertain world. For the most part we have attempted to manage uncertainty, even in the legal system. It is implicit in the jury system of UK law that we can never be sure of a person's innocence. No one leaves court pronounced innocent they are merely not guilty. As Bernstein says, 'when our world was created, nobody remembered to include certainty. *We are never certain; we are always ignorant to some degree.*'[36]

Uncertainty then, is not the enemy and certainty is a human construction that conceals a nasty surprise. In recognising this, writers such as Galbraith, van Creveld, Klein and Schmitt see managing uncertainty as the answer, possibly mitigating it by dividing it, but not removing it.

A more radical solution to the problems of uncertainty would be not to try and command in the conventional sense, but draw on examples from nature. Boid Theory is a biologically based model of complex systems developed by Craig Reynolds.[37] It attempts to understand how birds can flock together and execute seemingly complex actions with limited command structures. Reynolds's solution is that the birds follow very few, simple rules; (1) avoid other objects, (2) keep station with others, (3) head for the centre of mass of the flock. This is the basis of complex computer animation techniques that produce the stampeding dinosaurs in the film *Jurassic Park*, without hopelessly long programming requirements. This theory demonstrates that simplicity can deal with complex problems very effectively without large information requirements at high (in this case flock) level.

A further step in utilising biological models is the use of ant behaviour to solve complex problems. Software can be used to mimic ant colony behaviour[38] and is termed agent based modelling. Collectively ants achieve the most efficient solution to the problem of collecting food around their nest. At the start the optimal solution is uncertain and no ant has the information to say which is the fastest route to the food. As the ants search for food, they lay a pheromone trail and thus the first back will be the first to lay a trail that others can follow. Resources will then be applied to the most efficient solution, not by central guidance, but by individuals following simple rules. The situation is constantly changing, as food runs out and routes become blocked. Rather than waiting for redirection individuals can react to change immediately.

Modelling based on this technique can have counter intuitive results. South West Airlines modelled their cargo operations using agent-based models. Key elements represented were aircraft, packages and handlers. The traditional method of moving cargo to its ultimate destination was to follow the most direct route of airports, unloading and loading packages at each airport. It was discovered that if the packages were simply left on an aircraft they would eventually turn up at the destination, but in a faster time than the old method. This allowed South West to improve their response time without increasing their resource base. The time the package spends in transit is analogous with the time hierarchical structures take to make decisions. South West have flattened their structure, there is only one decision/action point – which plane to put the package on!

Agent based modelling has also had inputs into the modelling of commercial supply chains. Work by Strader, Lin and Shaw looked at the consequences of the sub division of tasks advocated by Galbraith and van Creveld. They suggested that,

> coordination in the past was facilitated through large hierarchical organisational structures. Today, large hierarchies are separating into smaller more specialised companies where co-ordination cannot be mandated. This has created a situation where uncoordinated inter organisational business processes result in unacceptable overall organisation performance, even if individual business units are operating efficiently.[39]

If this sort of comment seems out of place when discussing military organisations it should be compared with an assessment of army/air force co-operation in World War II,

> The general impression left on the mission in respect of the Burma front was that the Army was fighting one war and the Air Force another and that in consequence much precious effort was going to waste.[40]

Strader, Lin and Shaw[41] conclude that the management of uncertainty in supply chains is a trade off between cycle time, inventory and information. Until the 'information techno-logy' revolution, information had been expensive to acquire and interpret, so resources were held in lieu of knowledge. This is the logic of Just in Time production techniques, but the key is the quality of the information.

It is debatable as to what extent the quality of the information can be controlled in a military environment, remembering that an enemy will be actively attempting to degrade the quality of the information available.

Additionally, different production processes have differing lead times due to the nature of the product and the information required. The two most pertinent to this discussion are Make to Order (MTO) and Assemble to Order (ATO).[42]

MTO is a technique that produces custom products and is driven by customer orders, no components are held and thus it has long lead times.

ATO requires the holding of individual components in order to meet the requirement for custom orders but allows late differentiation. The uncertainty here is what the customer will order. In both cases this is unknown, but the crucial difference is that of time, if you have time it is more resource efficient to use MTO.

So far we have seen that uncertainty is the central military problem. Uncertainty is quantified as the gap between what we know and what we need to known in order to complete the task. Information is required but it is not discrete and can be best described as having two main attributes, quality and value. In approaching uncertainty organisations, civil or military, can adopt several strategies. The most successful approaches attempt to manage uncertainty by devolving decision-making and resources to lower levels.

Additionally, simplicity approaches such as those model-led by Boid Theory could offer an alternative approach to overcoming the problem of uncertainty.

Finally, the emergence of new information technology and approaches suggest that information handling techniques could allow the replacement of older systems based on resource holding to reduce uncertainty.

HISTORICAL EXAMPLES

Military commanders have, as Knight did, attempted to distinguish between what is literally uncertain and what is a quantifiable risk. Field Marshal Erwin Rommel would have agreed with Knight's proposition that man cannot predict the future, with one exception, 'The only time that a commander can calculate the course of a battle in advance is when his forces are so superior to the enemy's that his victory is self evident from the start.'[43] Rommel, with strong self-confidence, ignored the possibility of poorly led superior forces being defeated, but would have include the commander's attributes as part of the composition of a superior force.

Rommel distinguished between uncertainty and a quantifiable risk in terms of the outcome, defining the two options as, 'operational and tactical boldness and a military gamble'.[44] Rommel saw that there were no difference in the chance of success, but that the consequences of failure would be mitigated,

> A bold operation is one which has no more chance of success but which, in the case of failure, leaves one with

12

sufficient forces in hand to cope with any situation. A gamble, on the other hand, is an operation which can lead to victory or to the destruction of one's own forces.[45]

So Rommel's views coincided with Knight's proposition that the future is 'entirely unique' and invokes risk management techniques to mitigate the consequence of failure.

To Rommel reconnaissance (the gathering of information) was important, but it was the speed of the information flow and the commander's decision time that was important, 'Speed of reaction in Command decisions decide the battle.'[46] This had two outcomes affecting his processing of information.

First, he would slow the enemy via deception, to make 'the enemy commander uncertain and compel him to move with hesitation.'

Second, Rommel would not move with hesitation, even when lacking in firm information. His conduct of the Battle of Gazala (May–June 1942), demonstrated an ability to manage uncertainty that could be described by Schmitt and Klein's strategy of gripping the enemy to reduce one's own uncertainty. The battle saw many setbacks for Rommel but,

> Although often as ignorant as his opponents of exactly who was where or doing what, he nevertheless always gave the impression of being in control, of deciding with great rapidity, of drawing events in his wake…[47]

In a manner he was creating his own certainty, which he made into reality. He was not creating an abstract plan like Arrow's commanding general, but a series of short-term plans based on the information he had. Thus the surprises of the battle, such as the Allies' new Grant tank and the spirited Free French defence of Bir Hacheim, were overcome by rapid decision-making not more information.

The US Armed Forces in Vietnam demonstrated the opposite to this type of command. Their approach was characterised by van Creveld as one of 'information pathology'.[48] In contrast to Rommel's requirement for speed, intelligence sources even at divisional level were often useless due to, 'the chronic inability of the system to transmit information in time…'[49]

The result of the breakdown in intelligence/communications structures was that increasing amounts of time were required to plan and execute operations. By 1967 operations were taking between four and six months to plan, although the relief of Khe Sanh from 30 miles away only took two months to plan![50]

For the Americans in Vietnam, data and not information, had become the guide, just as the Army measured success via the body count, the US Air Force counted destroyed trucks.[51] This reflected the fact that those directing the war were more comfortable with a managerial-technical approach,[52] than one that advocated the need for rapid decision-making.

The chessboard analogy invoked earlier offers a link from the past to how some see future operations conducted. Some would argue that nineteenth century commanders often had full view of the battlefield in a manner that a twenty-first century commander will have on his screen. At Salamanca (1812), Wellington would be cast as the master, seizing the fleeting opportunity to checkmate Marshal Marmont as he over extended his line. Transferred to the twenty-first century, the commander merely waits for that key moment to appear on his screen before deploying his force for the decisive act. War is thus simplified, facts are known and all that is awaited is the moment to strike.

The contest at Salamanca was not a contest between master and novice. As Lawford and Young point out, Marmont was a skilled and experienced officer. Where he failed was in his judgement, which he based on Wellington's previous actions. He perceived Wellington to be a defensive general, 'who feared to conduct an offensive battle'.[53] Marmont's surprise at Wellington's assault confirms Knight's caution over predicting the future. Both commanders viewed the same information, but came to different judgements. This is not at all unusual and underlines the fact that success requires judgement, not necessarily more information.

It could be argued that Wellington's victory would have been more crushing had his chain of command not been of the chessboard type. From receiving the first indication that the French were over extending at 1415 hours, to personally ordering the 3rd Division to advance, a period of one and a half hours passed. It was not until 1645 that the 3rd Division clashed with the French. The advance of the 3rd Division had taken about an hour, so the majority of the time was spent observing, deciding and disseminating orders.[54]

The chessboard model is dependent on speed of decision making, Wellington had exclaimed 'My God, that will do' at 1445 but he could not act on his realisation for another hour. Proponents of the RMA argue that Wellington's intent could have been distributed to his commanders in seconds, but then so could Mamont's decision, which, based on the same picture, was fatally wrong.

Mamont mistakenly viewed Wellington as an 'asymmetrical' general, who preferred to not lose, rather than win. This type of approach seems to reflect naval tactics in the late eighteenth/early nineteenth century, until the advent of Admiral Lord Nelson. Naval actions of the eighteenth century were often indecisive and it was only at the end of the century that the battles at the Saintes (1782) and Camperdown (1797) indicated that there was a way to produce a decisive result. The tactic was to attack in line abreast across the enemy fleet rather than drawing along side. This would also steal the opponent's wind and allow for their destruction piece by piece. The old tactic supported the asymmetry theory. If the battle was going badly, then the commander could break and sail away downwind, with no danger of losing, but little chance of achieving decisive victory. This new tactic was considerably more risky as the guns of a ship's broadside could not be brought to bear until the last minute and it committed the fleet to fighting as there could be no breaking away. Trafalgar was to be the proving ground of this new tactic on the open sea.

Arguably Nelson had used this tactic at Copenhagen (1801) and the Nile (1798), but these battles were both fought in confined waters. From a strategic point of view the French fleet had to be defeated in order to relieve pressure on the British Navy, which was fully occupied blockading the French, to prevent them from invading Britain. Fortunately, the Franco-Spanish Combined Fleet commander, Villeneuve, was definitely a man who tried not to lose rather than win. In his mind the 'possible losses' (defeat of the Combined Fleet) certainly 'loomed larger than the gains' (control of the Channel).[55] This did not sit well with his supreme commander, Napoleon, whose strategy rested on Villeneuve gaining control of the Channel. Napoleon was to call Villeneuve 'a psychological if not moral coward'.[56]

Nelson was also aided by the introduction of the improved Popham codebook for signalling which allowed Nelson to control his fleet while conducting 'novel' manoeuvres. This codebook, by abbreviating common signals, allowed the rapid transmission of the commander's intent. To advocates of the latest RMA this could be the advantage that Nelson (Trafalgar, 1805) had over Wellington (Salamanca, 1812), his intent could be rapidly disseminated throughout the fleet. Nelson greatest asset was his style of command by which before the battle he ensured that his commanders understood his intent.

The value of signalling at Trafalgar is probably debatable, but it highlights how a commander can communicate his intent. Once battle was joined Nelson faced de facto decentralised command, leaving his band of brothers to complete his plan. As masts and cross spars began to fall the chances of effective signalling must have disappeared. Nelson's most famous signal of the day illustrates the problems any commander will have with any form of communication. Nelson had to modify his chosen phrase from 'England confides' to 'England expects' in order to conform more easily with the Popham signal book.[57] The former phrase expresses confidence in one's subordinates the later is more directive. This shows how a transmission medium can distort the message. The signal was arguably unnecessary as it conveyed no battle instructions and mis-conveyed his meaning, but the communications channel was, for a period, dedicated to its passage.

The signal probably had little impact on the battle, but in later years sowed the seed for a distinctively non-Nelsonian style of command, that saw obedience to orders as the key to success.[58] The conduct of the battle showed that Nelson had already communicated his intent to his subordinates, signalling was for minor detail and *bon mots*. Nelson managed the uncertainty of battle by ensuring his subordinates were in his mind before the battle. He was distributing the uncertainty and explained to his officers 'As for the handling of the ships when the smoke of guns obscured all signals, no captain could go wrong if he placed his ship along side that of the enemy.'[59]

Ultimately though, it was the difference in attitudes between commanders that probably led to the decisive victory. Both sides needed a decisive victory, but Villeneuve proved to be the perfect example of man's tendency toward asymmetry in decision-making. In trying not to lose he was comprehensively beaten.

Asymmetry can occur on the same side between fellow commanders where it can lead to defeat or indecisive victory. A modern example would be General Omar N. Bradley's defence of his failure to close the Falaise Gap during the 1944 breakout from Normandy. The German 7th Army could only exit Normandy via a gap between Argentan and Falaise. With the Canadians moving slowly south, General George S. Patton saw an opportunity to link up with them by driving north from Argentan. Concerned about meeting head on Bradley, Patton's superior, refused him permission. Bradley defended his action by stating that he preferred 'a solid shoulder at Argentan to a broken neck at Falaise'.[60]

Bradley was correct to say that this was not part of Montgomery's plan, but the closure of the gap would have prevented the withdrawal of 20,000 German troops. Like the doctors mentioned by Bernstein, when presented with a new option Bradley preferred to stay with his original plan despite the chance to increase Allied gains. Patton presented Bradley with new information that was not utilised to its best effect, demonstrating that information is, by itself, limited in value.

Van Creveld identified that the collection and dissemination of information can debilitate the command function. The shooting down of an Iranian airbus by the cruiser USS *Vincennes* in 1988 suggests that acting on new information can be just as dangerous as not acting. Klein suggests[61] that the crew of the *Vincennes* used six pieces of information to decide that the airliner was hostile. Four pieces of information suggested that the airliner was behaving abnormally (e.g. not on schedule, off course etc). The last two pieces of information were later seen to be incorrect, but suggested that the plane was hostile. When the decision was made the aircraft was thought to be descending and emitting an Identification Friend or Foe (IFF) signal comparable with an Iranian F-14 jet fighter. The guided missile frigate USS *Sides* also saw the same aircraft, but concluded that it was not hostile.

The difference in interpretation comes down to the last two pieces of information, which the crew of the *Vincennes* interpreted differently. Klein suggests that the IFF signal was due to human error and the descending track due to confusing the airbus with another aircraft.[62] Other theories based on expectations and biases following previous incidents were also suggested and put forward by the official report.

What stands out is that this error was based on six pieces of information. Two of these pieces of information were wrong and had not survived contact with the human machine interface.

More information than the six pieces they possessed would not have clarified the situation within the timeframe the crew had to decide (before a hostile track got within missile range). They had the information that was of *value*, but did not know how they could improve the *quality*. The only people that could improve the quality of the information were the crew, but they did not know they were in error. Information could not make the decision and proved to be poor at supporting the decision of the commander.

In contrast to the *Vincennes*, the US commanders in Somalia had a surfeit of information but proved unable to act faster than the local militias. In this case Boid theory could provide the answer as to why the Ranger Task Force was overwhelmed in 1993. In terms of command and control infrastructure, the Americans seemingly had the upper hand. In addition to unit radios, there were several airborne command platforms and links back to a centralised operations room. As the operation began to go wrong, following the loss of a Black Hawk helicopter to a rocket-propelled grenade, the Americans failed to coordinate the actions of their forces.

In contrast the Somalis rapidly countered all American moves. It could be argued that they achieved this by following three simple rules; (1) move to the sound of gunfire, (2) shoot any American seen, (3) tell someone else to join in. Using these rules their tempo was considerably higher than the Americans. That the Somalis did not consciously decide to organise in this manner does not matter. Their behaviour demonstrates how a distributed group can work to a common goal with no overall common picture. The danger of this approach is demonstrated by the ease with which fishermen catch shoals of tuna in funnel nets – with no big picture Boid flocks contain the seeds of their own destruction.

The last method of dealing with uncertainty mentioned above was the different approaches to supply chain modelling. Assemble to Order (ATO) processes seem to describe the provision of Close Air Support (CAS) or artillery support to the close battle. In both cases the target effect is the product. The 'customer' does not know where or what he requires until the last moment and so requires a system that is very responsive. As a consequence of this, stocks are held in terms of aircraft in cab ranks or guns placed in direct support. Due to the short timeframe, stocks have to be held as the method of managing the uncertainty. Information cannot replace stocks, as the information does not yet exist.

This type of support is unpopular with the owners of the assets, as they view stockholding as an uneconomical use of the assets. ATO is only possible if communication technology allows the target information to be passed in a rapid and timely manner to a system that can produce a rapid effect. These technological problems were initially grappled with in World War II in both the European[63] and Far East theatres, but the underlying tension was the resource issue that was in itself generated by attempts to manage uncertainty.

Before the advent of reliable radio communications, the provision of indirect fire had to be Make to Stock (MTS). With no timely input from the customer, all that could be achieved was a production line churning out fire, which was not responsive to the needs of the soldiers involved in the close battle. This is the nature of indirect artillery support in World War I. The week-long bombardment prior to the Somme in 1916 was artillery support based on a Make to Stock approach. This was not due to a lack of imagination on the part of the planners, but simply because initially there was no way to provide a responsive system.

Writing about the Battle of Neuve Chapelle in 1915, Alanbrooke, then a Staff Captain, noted that despite numerous individual telephone lines there was no telephone exchange, leading to chaos.[64] Even innovations such as creeping and hurricane barrages were variations on a theme trying to make the guns more responsive. By the end of the war artillery spotting used balloons and aircraft, but this was also a third party method.[65]

Until the technology was developed to enable the soldier requiring the fire to direct it, fire support would remain MTS.

MTS is essentially forward planning and trying to predict what your requirements will be. Schmitt and Klein would recognise this as a method for managing uncertainty, but 1914–18 demonstrated its limits.

The aim of this chapter is not to debate whether an RMA is occurring. It is assumed that the digitisation process underway in the USA and Britain will continue with the aim of producing 'a marriage of systems that collect, process and communicate information with those that apply force'.[66] The US military is particularly keen on driving this process forward and is developing a vision of 'network centric' warfare that will, 'allow war fighters to take full advantage of all available information and bring all available assets to bear in a rapid and flexible manner'.[67]

In both of these definitions, and most others, information is seen as key to improving the efficiency of warfare. From the previous section there appears to be two issues that information influences, resources and command decision. In order to allow a commander to make the correct decision, or for resources to be applied, uncertainty must be managed. The next section will analyse the promised RMA, in order to see if it will be able to improve the management of uncertainty in these areas.

Much of the drive behind the recent RMA is to increase the speed with which the OODA[68] loop can be transited – this may result in a highly centralised organisation. As you do not control the enemy, the tempo of the 'observe' part of the loop is not within your control, only the span of observation can be extended. The remainder of the loop is dependent on an individual's speed of reaction to stimuli and then on the system's ability to communicate the individual's intent to whatever will implement the decision. This will be achieved by information dominance that will provide situational awareness, dissipating the fog of war by the use of a common view.[69] Just as with the FIST trials, subordinates will be able to anticipate commands increasing the tempo of the operation.

Additionally, 'real time' sensor to shooter architecture could eliminate the need for command levels[70] and certain capabilities (e.g. indirect fire) would not have to exist at each level. These last points mean that command could be simplified and logistical burdens could be reduced as firepower is centralised. Centralisation seems to be a likely outcome of such a process as the centre will hold 'the big picture' and, importantly, control the resources.

The foregoing suggests a return to the nineteenth century style of command, but the issue of certainty will still remain. Commanders will be working by proxy, not using their own direct experience and this presents several problems.

First, commanders will still need to interpret the facts presented but, as seen previously, this is not easy. Smith[71] suggests that commanders produce three reactions when attempting to understand a picture; (1) recognition and correct analysis, (2) recognition but incorrect analysis, (3) failure to recognise. These reactions will not change, as it is a mental process conducted by the commander.

Second, commanders are on a continuum and they need to recognise when they have the, 'minimal essential information to decide on a course of action.'[72] As information is collected time is passing and therefore the ability to change the situation is constantly reducing, until a crisis point arrives when no action is possible. The logic of this is that the problems of uncertainty, such as asymmetry, will not be overcome by technology, but by improving individual judgement.

Commanders will remain key to a digitised system, but the process and structure may degrade commander's ability to make decisions. Stanley-Mitchell suggests that skills-fade will follow automisation of tasks and indecisive leaders will be overwhelmed both by the volume of data and by the cognitive dissonance between computer 'truth' and their actual experience.[73]

An increasingly centralised system will remove the levels of command where people gain their experience and also tie commanders to their command post, so their feel for the battle would be second hand. Importantly, as Schmitt and Klein note, 'Decisions will be made under the same level of uncertainty as before because decision cycles will accelerate. Commanders will sacrifice certainty for tempo.'[74] The implication of this is that less able commanders will be required to deal with the same level of uncertainty as their predecessors.

If the RMA is simply doing what we do now, but faster, then the true benefits of the information technology will not be realised. If there is to be no doctrinal or philosophical shift some would argue that this is not a RMA at all. Sir Michael Howard argued that the Royal Navy failed to make the doctrinal shift, from the direct fire that won Trafalgar to the indirect fire that would dominate the twentieth century, quickly enough. The doctrine of cannonade was seen as essential, 'the immortal memory of Horatio Nelson ensured the survival of the doctrine in the Royal Navy long after technological developments made it impossible'.[75] If an RMA is occurring then there must be a major doctrinal shift.

Evidence from the US Army 1997 Divisional Advanced Warfighting Experiment suggests that initial use of the latest technology is not revolutionary. Commanders tended to use the data made available to 'feed their artillery more targets at which to shoot'.[76] Targets were engaged at long range and this could be seen as the start of the idea that the Deep Battle will become the decisive engagement. This sits with the idea of not losing rather than winning. The concept of deep battle is often couched in terms of increasing the certainty of outcome, the phrase 'create the conditions for...' being very common. Manoeuvre in this logic is to support fire and this seems to be the more cautious approach – a high tech version of World War I barrages designed to allow the infantry to mop up.

The more serious problem with visions that advocate creating greater certainty as the key to success is that they may stifle initiative and creativity. A comment following the 1997 Warfighting Experiment is illustrative of the point, 'even if I do stray, my boss will see the error almost immediately and correct it'.[77] The individual quoted has surrendered his initiative to his superior, he is no better than one of Frederick the Great's automatons, or one of Wellington's divisions. This is not someone that will seize a fleeting opportunity, for him there is no uncertainty.

Uncertainty has been centralised again and thus rests with the one central commander. This is also because the central commander is best placed to view the whole picture, unlike the tank commander who will be viewing events on a far smaller screen. This means that chances to micro-manage will multiply, reinforcing command by direction approaches, which seem to be the antithesis of the approach advocated by van Creveld.

There is also likely to be a bias towards investment in surveillance and command assets. Vice Admiral Jeremy Blackham, Deputy Chief of the Defence Staff (Equipment Capability) stated in October 2000 that Integrated Surveillance Targeting and Reconnaissance (ISTAR) assets would receive an 'increasing proportion of our research and development and of our capability acquisition budget....quite possibly at the expense of some of our platforms'.[78] As budgets are limited this means that the ability to manoeuvre will not develop at the same pace as ISTAR and the amount of ammunition that can be bought will be limited.

The technology is neutral, but currently the emphasis seems to be on keeping the same structure and just going faster. A step change in thinking would be to draw lessons from Rommel and Somalia. The new technology could be used to ride the chaos of the battlefield by communicating intent and allowing access to resources. This approach draws on an idea put forward by Lanir and others,[79] who suggest that when faced by a chaotic situation, either the old order can be restored or, utilising creativity, a new order is established.

The old order takes us no further; Rommel at Gazala was utilising creativity to create a new order. This approach requires the use of short-term plans, modification and the ability to decide quickly on partial information. Until the end state is reached the situation will remain uncertain, but this is required if creativity is to be allowed to bloom.

To use Rommel as an example of successful warfighting is of course limited by the fact that he was ultimately defeated. He was defeated because of the second factor that the RMA must consider – resources. Resources can be used to manage uncertainty, but resources will always be limited. Freedman contends, 'The RMA may combine systems that collect, process and communicate information with those that apply military force, but the most important of these **is** the military force.'[80] Any increase in stand off weapons and precision attack will centralise the resource of military force that, as Freedman points out, is essential to warfighting. Without control of resources, individuals no longer have influence;

they can have no impact on the situation. As with command, the issue of resources will tend towards centralisation, reducing the flexibility of the system.

CONCLUSION

The first part of this chapter suggested that certainty could not be achieved. This is because no situation is ever an exact reproduction of past events. If, therefore, plans are based on a prediction of the future, extrapolated from the past, they will be flawed. Commanders are thus forced to work in an environment that is uncertain. There are many theoretical ways to manage the resulting uncertainty, but a common strand in the theoretical approaches is to break up problems into tasks that can be completed by self-contained sub-units. Other alternatives seem to offer sub-optimal solutions.

Information is the key requirement that allows commanders to make decisions. The structure of an organisation reflects how it will deal with the problem of providing the commander with information. It is now recognised that individuals do not react predictably to information. Individuals are often cautious in the face of new information, and tend to work from a few cues. Few people work through problems by weighing all the factors, deciding and then formulating a plan. This implies that more information is not required, unless it is recognised by the individual to be of value and have the requisite quality. The key factor in historic commanders' decision-making is their judgement, or their ability to identify the pattern.

If the RMA produces a chessboard style solution it will not be truly revolutionary. Centralisation of command and resources is a possible outcome of the current technical approach. This runs contrary to the optimal theoretical approaches to uncertainty outlined above, and to the successful approaches of the historical commanders reviewed.

The technology is, however, neutral and it is *how* it is used that will be crucial. Once the commander has made his judgement, the important factor is the speed and accuracy with which his intent can be communicated. His subordinates must then have the resources to enact the commander's intent. This should be the aim of any future RMA. This may require adapting traditional structures and training individuals to cope in chaotic environments. Uncertainty in such structures should not be resisted, but recognised as the creator of the fleeting opportunities that bring victory.

REFERENCES

1. *Chamber's Twentieth Century Dictionary* (1961) p.172.
2. Van Creveld, M. *Command in War* (Cambridge, MA: Harvard University Press 1985) pp.264–8.
3. Lawford J.P. and Young P. *Wellington's Masterpiece: The Battle and Campaign of Salamanca* (London: George Allen and Unwin 1973) p.281.
4. Schmitt and Klein G. 'Fighting in the Fog', *Marine Corps Gazette* (August 1996) p.62.
5. Howard M. *et al.* (eds.) *Clausewitz On War* (Washington DC: Library of Congress, 1996) p.144.
6. Liddell Hart, B.H. *The Other Side of the Hill* (London: Collins 1951) p.7.
7. Hayek F.A. *The Fatal Conceit (The Errors of Socialism)* (London: Routledge 1988) p.88.
8. Heaney S. *Beowulf* (London: Faber 1999) p.xvii.
9. Bernstein P.L. *Against the Gods – The Remarkable Story of Risk* (New York: Wiley 1998).
10. Ibid. p.220.
11. Quoted in Bernstein (note 9) p.221.
12. Fuller J.F.C. *The Foundations of the Science of War* (London: Hutchinson 1926) p.38.
13. In Fuller's defence he was also arguing for a more systematic approach to the study of military history. See Welch M. 'The Science of War: A discussion of JFC Fuller's shattering of British continuity', *Journal of the Society for Army Historical Research* 79 (2002) p.327.
14. Galbraith J. *Designing Complex Organisations* (Reading, MA: Addison-Wesley 1973) p.4.
15. Ibid.
16. Ibid. p.5.
17. Van Creveld (note 2) pp.268–9.
18. Ibid. pp.247–9.
19. Ibid. p.269. See also Macgregor D.A. *Breaking the Phalanx. A New Design for Landpower in the 21st Century* (Westport, CT: Praeger Press 1997) pp.1–2, for a description of the fate of the Macedonian phalanx in Thessaly.
20. Schmitt and Klein (note 4) p.63.
21. Ibid. p.64.
22. DLW *The 2015 Battlefield* (2001) p.11.
23. Schmitt and Klein (note 4) p.65.
24. Ibid. p.66.
25. In support of this they point out that radar has not reduced ship collisions and weather forecasts become more unreliable with more information.
26. Schmitt and Klein (note 4) p.69.
27. Darilek R. Perry W., Bracken, J., Gordon, J. and Nichiporuk, B. *Measures of Effectiveness for the Information Age Army* (Santa Monica, CA: RAND Corporation 2001).
28. Ibid. pp.12–19.
29. Bernstein (note 9) p.274.

30. Ibid. pp.278–9.
31. Clausewitz (note 5).
32. Ibid. p.140.
33. Ibid. p.144.
34. Arrow, 1992 quoted in Bernstein (note 9) p.203.
35. Ibid.
36. Ibid. p.203.
37. <www.red3d.com/cwr>.
38. Whiting R. 'Radical Simplicity Behaviour Change for Supply Chains', *Information Week*. 2 April 2001.
39. Strader T.J., Lin, F. and Shaw, M.J. 'Simulations of Order Fulfilment in Divergent Assembly Chains', *Journal of Artificial Societies and Social Simulation* Vol.1, No.2 (March 1998).
40. Report by Military Mission 220, 1944, p.7.
41. Strader *et al.* (note 39) p.12.
42. The last is Make to Stock (MTS).
43. Rommel Papers, quoted by Young, D. *Rommel* (London: Collins 1950) p.256.
44. Ibid. p.255.
45. Ibid.
46. Ibid, p.254.
47. Fraser D. *Knight's Cross: A Life of Field Marshal Erwin Rommel* (London: HarperCollins 1994) p.327.
48. Van Creveld (note 2) p.247.
49. Ibid.
50. Ibid. p.249.
51. Whitcomb D.D. 'Tonnage and Technology. Air Power on the Ho Chi Minh Trial', *Airpower History* Vol.44, No.1 (1997) pp.4–17.
52. Krepinevich A.F. *The Army and Vietnam* (Baltimore, MD: Johns Hopkins University Press 1984) p.35.
53. Young and Lawford(note 3) pp.284–5.
54. Ibid. pp.222–44.
55. *The World at War*, Channel 4, 24 March 2002.
56. Keegan J. *The Price of the Admiralty* (Harmondsworth: Penguin 1998) p.30.
57. Hibbert C. *Nelson. A Personal History* (Harmondsworth: Penguin 1994) p.366.
58. Andrew Gordon gives a good account of the decline of the British Navy command in the century after Trafalgar *The Rules of the Game. Jutland and British Navel Command* (London: John Murray 1996)
59. Hibbert (note 57) pp.359–60.
60. D'Este C. *Decision in Normandy* (Harmondsworth: Penguin 1983) p.430.
61. Klein G. *Sources of Power* (Boston, MA: MIT Press, 1998) p.82.
62. Ibid. p.84–6.
63. See Gooderson, I. *Air Power at the Battlefront: Allied Close Air Support in Europe 1943–45* (London and Portland, OR: Frank Cass 1998) for a discussion of CAS in Europe.
64. Fraser D. *Alanbrooke* (London: HarperCollins 1997) p.38.
65. Keegan J. (note 56) pp.341–2.
66. Freedman L. 'Britain and the Revolution in Military Affairs', *Defense Analysis*, Vol.14 No.1 (1988) p.55.
67. *Network Centric Warfare: Department of Defence Report to Congress*, July 2001, <www.dodccvp.org/NWC/NCW.report/start.htm>.
68. Observe Orientate Decide Act after Col. Boyd.
69. Stanley-Mitchell, E.A. 'Technology's Double Edged Sword: The Case of US Army Battlefield Digitisation', *Defense Analysis* Vol.17, No. 3 (2001) p.269.
70. Ibid. p.270.
71. Smith, K.B. 'Combat Information Flow', *Military Review* Vol.69 (April 1989) p.43.

72. Ibid. p.48.
73. Stanley-Mitchell (note 69) p.276.
74. Schmitt and Klein (note 4) p.66.
75. Howard M. 'War and Technology', *RUSI Journal* Vol.132, No.4 (1987) p.17.
76. Leonhard R.R. 'The Culture of Velocity', *Digital War* (Novato, CA: Presidio 1999) p.137.
77. Ibid.
78. Blackham J. 'The Apotheosis of 21st Century Warfare', *RUSI Journal* Vol.145, No.6 (Dec. 2000) p.67.
79. Lanir, Z., Fischhoff, B. and Johnson, S. 'Military Risk Taking: C^{3}I and the Cognitive Functions of Boldness in War', *Journal of Strategic Studies* Vol.11, No.1 (March 1988) p.101.
80. Freedman L. (note 66) p.56.

2

Reflections on 11 September 2001

JACK SPENCE
Royal College of Defence Studies

Never glad confident morning again
Robert Browning (1812–89)

Any commentary on the atrocities on 11 September 2001, designed to offer pertinent observations on the likely changes in the structure and process of international relations thereafter, must inevitably be cautious, both in tone and substance. The Taliban regime has fallen: an interim administration has been established in Kabul, but uncertainty remains about the time scale and resources required to bring the 'war' against terrorism to a successful conclusion.[1] Furthermore, there is still debate about how far the aims of the coalition should widen beyond the capture and/or assassination of Osama bin-Laden and the destruction of his organisation.

Yet on the evidence of what has happened so far, it is possible to raise at least seven questions. These are listed below and are by no means inclusive. There are, no doubt, many others deserving detailed examination, but these particular views are offered in the hope that what follows will at least provoke debate:

- First, the extent to which al-Qaeda represents a new and distinct form of terrorism;
- Second, the nature and scope of the so-called 'war' against terrorism and the degree to which it differs from earlier conflicts;
- Third, the impact of the campaign on the structure and process of interstate relations, especially with regard to the great powers;
- Fourth, the staying power of Western electorates, their capacity to cope with the unexpected in terms of the variety of threats, and the capabilities available to the terrorist;

- Fifth, the likely effect on domestic politics, especially with regard to intrusion into areas such as civil liberty – hitherto regarded as beyond the reach of legitimate interference by the state;
- Sixth, the implications of prolonged engagement against terrorism on Western attitudes to the persistence of the North–South divide; and what, if anything, can and should be done to narrow the gap, not simply in material terms, but also with respect to perceptions; and
- Finally, what the likely consequences of the 11 September attacks for the academic study of international relations will be.

All these issues overlap, as the following analysis demonstrates.

TERRORISM: OLD AND NEW

Terrorism – often described as the weapon of the weak – is a particular form of asymmetric warfare with which Western governments have had to contend in a variety of guises and contexts since 1945. It is rarely – if ever – mindless, but is rarely successful unless employed as a precursor to the launching of guerrilla or revolutionary war: 'a protracted struggle in which irregular military tactics are combined with psychological and political operations to produce a new ideological system or political structure'.[2]

In other words, the orthodox terrorist-cum-insurgent is Clausewitizian in outlook, concerned to use the weapons at his disposal – sabotage, assassination, intimidation, random attacks on government structures – to achieve a political objective such as the withdrawal of a colonial power or the liberation of a people from alien or repressive rule. Many examples come to mind: the Greek Civil War in the 1940s; Palestine in the same decade; Cyprus in the 1950s; Vietnam in the late 1950s and 1960s.

John Baylis:[3]

> distinguishes between two forms of terrorism often used simultaneously: coercive terrorism is designed to demoralise the population and weaken its confidence in the central authorities as well as to make an example of

29

selected victims. Disruptive terrorism is designed to discredit the government, advertise the movement and provoke the authorities into taking harsh, repressive counter-measures.

Both these definitions are helpful for an understanding of the al-Qaeda movement. Certainly, demoralisation of America's citizenry was a primary goal; the victims were selected with care, given that both the Twin Towers and the Pentagon were prime and potent symbols of the capitalist system and the military might of the US. The movement, too, gained massive publicity. Indeed, the destruction caused was paradoxically the result of the violent juxtaposition of two highly vulnerable instruments of global capitalism – the aeroplane and the skyscraper housing a representative sample of bankers and traders. No doubt, too, bin-Laden's strategy was designed to provoke 'repressive counter-measures'. Yet the American people, despite the initial shock, were not demoralised, and demonstrated a commitment to old-fashioned patriotism and retaliation in response to what was perceived to be profoundly evil.[4]

Yet there remains a significant difference between al-Qaeda and its terrorist predecessors. The motivations and political objectives of, for example, the Vietcong in Vietnam or the Algerian Army of National Liberation in Algeria were relatively limited, driven by a profound desire for national self-determination and the achievement of statehood. Once their goals were achieved, as members of the newly established governments they became in due course respectable, incorporated as they were into the society of states and rapidly socialised into the norms, conventions and rules of that society.[5]

This, incidentally, was equally true of South Africa's African National Congress, which, while condemned as a terrorist organisation in some quarters (notably the Thatcher government in the UK), became – following the transition from the apartheid regime – a solid and well-respected participant in world affairs.

By contrast, al-Qaeda seems to be moved by ambitions which are unlimited in scope and substance. Fuelled by unadulterated hatred and sheer rage at the evident inequalities between the rich North and a poverty-stricken South, bin-Laden appears to want nothing less than a radical restructuring of international society. True, his movement has focused on the plight of the Palestinians and the American presence in the sacred Islamic heartland of Saudi Arabia. But these are objectives subordinate to an overwhelming desire to radicalise Islam into a confrontation with a Western world regarded as corrupt, secular in the extreme, arrogant and indifferent to the claims of the poor and the oppressed. His objectives, therefore, are not narrowly defined, but global in their ramifications.[6]

And the question remains: Can terrorism thus defined follow the pattern set by its counterparts in the past and move to the next stage, namely revolutionary warfare on a global scale? Clearly not, but this caveat does not exclude fomenting revolutionary struggles in, for example, Middle Eastern regimes which appear to be the creatures of Western imperialism.

Nor can we rule out the sowing of disaffection in Western states with large Muslim minorities. In the UK, for example, radical religious leaders have incited young Muslims to enrol in the Taliban struggle. And we should in this context bear in mind the impact of the images shown by the Western media, perceived as evidence of luxury, decadence and sexual licence, which are beamed into Muslim homes across the globe on a daily basis.

HEARTS AND MINDS

On the assumption, then, that bin-Laden has embarked on a global strategy, Western governments have little choice but to fashion an appropriate long-term 'counter-insurgency' response. Traditionally, 'winning hearts and minds' has involved a combination of political and military measures designed to contain violence and restore a degree of security to those most harmed by 'disruptive terrorism'. At the same time, political structures designed to combat the ideological appeal of opponents have to be devised. (One notable success in this context was the British campaign in the 1950s in Malaya against the Communist insurgency.)

In many cases, however, the revolutionaries won; colonial or alien presences, in particular, had little alternative but to negotiate face-saving withdrawals, as in the Paris peace talks in the 1970s between Henry Kissinger and his Vietnamese opponents.

But at least in all these cases diplomatic negotiation between victor and vanquished was possible, based on well-tried techniques and modes of political communication. Diplomacy, however, even in the limited context of compelling the Taliban to surrender bin-Laden, failed. Indeed it is improbable that bin-Laden and his followers will be susceptible to diplomacy of any kind – whether by orthodox means of persuasion and negotiation, or the more coercive variety practised before the bombing campaign began.

The task facing the West after 11 September 2001 is, therefore, infinitely more difficult: it is to undercut the appeal of fundamental extremism on a global basis, and at the same time eradicate the source of the threat. The latter, however, is not confined to one country: it has supporters throughout the Middle East and elsewhere, for whom a narrowly-based education in, for example, the religious schools in Pakistan provides a fertile breeding-ground for hostility to the West in general and the US in particular. Eliminating bases and political structures sympathetic to bin-Laden in Afghanistan and even capturing or assassinating him – two war aims of the current campaign – represent only the beginning of what most observers and practitioners agree will be a 'long haul' (Tony Blair's phrase).

What, in effect, is envisaged is a campaign *in perpetua*, the conclusion to which the West can never be certain of, and from which there is no obvious 'exit strategy'. In this context, winning hearts and minds involves not only gaining the support of Third World states (especially those in the Arab world), but also the support of their peoples.

And there is a further difficulty: many of the governments – especially in the Middle East – are authoritarian, a factor complicating Western political strategies. Does the West continue to prop them up, hoping that economic liberalisation will lead to political liberation of a democratic variety over the long term? What happens if popular revolt happens first? How will the West react? Thus, a 'hearts and minds' strategy is a formidable undertaking, given the hostility towards what is perceived to be the overweening arrogance of a global 'McDonald's' culture.

And even if bin-Laden and his associates are captured and put on trial, how could a judicial venue be found that is acceptable to the international community as a whole? Where could a sentence be served without endangering the security of the state concerned? Nor would straightforward assassination end the struggle. Either outcome would brand bin-Laden as a glorious martyr to his cause, and sow the seeds for further rebellion.

An alternative strategy might be to keep bin-Laden (and any potential successors) on the run, denying him and his movement the time, security and means to plan further major terrorist attacks. After all, the planning of the 11 September atrocity was done over a long period, during which bin-Laden was relatively safe from major harassment.[7]

Indeed, a seemingly never-ending struggle to win hearts and minds on a global scale – an open-ended objective by any standards – is at odds with the traditional Clausewitizian prescription for the successful conduct of war: a precise objective, political will, and a clear exit strategy. And this difficulty will be compounded if the coalition's war aims widen beyond the destruction of al-Qaeda and the current Afghanistan regime to include military intervention elsewhere, in support of a global anti-terrorist campaign. The political will of the UK and the US is currently not in question; but can it be sustained uniformly in the coalition as a whole over the long term?

And calculations of military means and objectives are further complicated by the sheer uncertainty of when and how the enemy will strike next. In other words, the intermittent nature of the struggle over the long term might well make very difficult the sustaining of popular support – an essential *domestic* imperative in any 'hearts and minds' campaign.

This 'war' – if that is what the West is in for – will be unlike any other conflict in the past: the battle lines will be hazy, and its key operatives will be intelligence gatherers and financial institutions, working quietly and surreptitiously to support any military action required. Furthermore, all this will require a high degree of multilateral cooperation which will be uneven in scope and substance.

Thus this 'war' is profoundly different from either the Cold War or its more orthodox predecessors. In such instances the disposition of the enemy was abundantly clear. Victory for one side or the other was certain in conventional conflicts, and military strategies could be devised with some assurance of ultimate success. Even in the Cold War, despite its ideological overtones, techniques of crisis management were created. These involved, *inter alia,* arms control measures to reinforce nuclear deterrence in the recognition by both the West and the Soviet Union that each had a common interest in avoiding mutual assured destruction and preserving an admittedly precarious global order, however damaged that may have been by proxy conflicts in the Third World. None of these advantages pertain in the war against terrorism. (It is worth noting that the IRA often issued coded warnings before bomb explosions – an interesting example of primitive crisis management!)

Short of a fundamental restructuring of the current international system involving massive rehabilitation of the world's most depressed societies, hostility to the West – whether deserved or undeserved – will persist. To be fair, Tony Blair, in his October 2001 speech to the Labour Party Conference, gave eloquent voice to the idea of a 'new international community' dedicated to eradicating poverty and 'healing the scars of Africa'. But this excited relatively little comment at home and abroad, given the media's obsession with the unfolding crisis in Afghanistan. Nor is it clear that the US has much appetite for devoting considerable resources to long-term state and nation building in the Third World.

The conventional wisdom that overseas trade and investment, coupled with the creation of an effective enabling political environment, is the only long-term solution to meeting the needs and aspirations of the 'wretched of the earth' will not be lightly jettisoned in favour of some elaborately structured mechanism of redistribution of global wealth. The best we can hope for is summed up by a leader in the *Financial Times*.[8]

> The ills of the world's poor result from too little globalisation, not too much. Their continued marginalisation can only perpetuate deprivation and a sense of injustice. But the extraordinary spirit of international co-operation engendered by last month's atrocities may just offer a gleam of hope. It has created new possibilities for binding the world closer together,

economically as well as diplomatically...there is a moment to be seized now...

Yet governments, by definition, operate pragmatically, coping with particular circumstances as best they can, and subject to a host of domestic and external constraints. As Hedley Bull, echoing Michael Oakeshott's famous metaphor,[9] argued:[10]

> The decisions of governments on matters of peace and war do not always reflect that careful weighing of long range considerations, or mastery of the course of events; the questions which strike the historian of these decisions a generation afterwards as important appear crudely answered or, more often, not even asked. The governments appear to him, to stumble about, groping half blind, to [be too] preoccupied with surviving from day to day even to perceive the direction in which they are heading, let alone steer away from it.

No doubt it could be argued that the present crisis, and especially its long-term implications, requires a sea change in attitudes and policy if Western governments are to cope effectively with global terrorism. Yet sustaining a campaign of this magnitude will require perpetual vigilance; considerable resources (including revising the current downward trend in defence expenditure); a massive improvement in intelligence gathering and assessment; a heightened degree of regional and international cooperation; and a reinvigoration of diplomatic efforts to reach a settlement of the Palestinian issue.

This is a formidable agenda which is likely to concentrate the minds of Western leaders to the exclusion – at least for the foreseeable future – of major reform of the global system sufficient to satisfy the needs of the world's deprived. True, the current crisis may well prompt an acceleration of measures designed to liberalise trade in favour of Third World products via the mechanism of the World Trade Organisation. True, renewed efforts in the Middle East to satisfy the conflicting aspirations of Israelis and Palestinians would be a major breakthrough, but it would involve compromises on both sides, leaving open the prospect of renewed violence from those movements which appear bent on Israel's destruction.

The point to stress here is that these solutions are partial, perhaps faltering, steps towards the realisation over the long term of Tony Blair's vision of a 'new international community' acting decisively to ameliorate the lot of the world's needy and dispossessed. This is at best a long-term aspiration; in the short- to medium-term, given the vengeful quality of modern terrorism, it is highly improbable that such achievements would eliminate completely the threat of further haphazard destruction.

This suggests that for a long time to come, Western governments will have to remain on an unconventional war footing, with all that implies for the formulation and conduct of foreign policy – an issue to which we now turn.

THE IMPACT OF THE TERRORIST THREAT ON THE STRUCTURE AND PROCESS OF INTERNATIONAL RELATIONS

Change in this area is already apparent. The Bush administration had previously displayed a commitment to unilateralism, as reflected in the refusal to ratify the Kyoto Climate Agreement; its persistence in proceeding with the National Missile Defence Programme; and a reluctance to sign the Biochemical Treaty. This was a reaction to what he and his advisers regarded as President Clinton's woolly multi-lateralism. However, in the light of recent events, the Bush administration has abandoned the unilateralist stance.

Instead, the US has engaged diplomatically on a variety of fronts in order to muster a credible 'coalition of the willing'. Pakistan, hitherto ostracised, has been assiduously courted; closer links have been forged with Russia and China (both of whom have cause to resist Muslim fanaticism within their jurisdiction, and who clearly expect in return a softening of Western attitudes on human rights issues in both states); new allies have been sought in the Middle East; noises have been made about the aspirations to statehood of Palestinians.

The United Nations (UN) has been paid some of the dues owed to it, if only because the US has been forced to recognise that the organisation will have to play a central role in the reconstruction of Afghanistan.

In this context the 'special relationship' with Britain has been revived, with Tony Blair acting – with varying degrees of success – as Bush's diplomatic cheerleader. Here he is following in the footsteps of his predecessors, in attempting to give renewed scope and substance to the notion of 'Britain punching above its weight' in international affairs. Again, in line with traditional post-war British foreign policy, Tony Blair is bent on acting as Europe's conduit to the US and vice versa. Moreover, he has one considerable advantage – well trained, adaptable and battle-hardened armed services at his disposal.

What we are witnessing, therefore, is a resurgence of state interest in combating what is perceived to be a fearful threat to security. International and regional organisations have been sidelined in favour of the creation of an informal alliance in which the US, as the salient superpower with enormous capability, is the lead state. The US invoked Article 51 of the Charter to justify its campaign against the Taliban, and successfully sought Security Council support for resolutions condemning terrorism, while the North Atlantic Treaty Organisation (NATO) passed an Article 5 resolution mandating members to engage in collective defence actions.

Yet both the UN and the European Union (EU) have taken a back seat in the handling of the crisis. The latter has concerned itself with soft security issues, devising regulations for humanitarian aid; illegal immigration; money laundering; more efficient extradition arrangements; and even – in the UK at least – proposals to detain suspected terrorists indefinitely. This is important as part of a Europe-wide programme for dealing with global terrorism.

However, what is striking is the emergence of three key states – the UK, France and Germany – as the major players, operating, to all intents and purposes, outside the EU framework. All three have committed forces in support of US efforts, with Blair in particular demonstrating a commitment to stand 'shoulder to shoulder' with the US. What this suggests is that intergovernmental crisis management is the preferred mode of international cooperation. This is a major departure from EU orthodoxy, which, via the Maastricht treaty, was meant to provide 'a blueprint for a common foreign and security policy...in which all members are supposed to move in step'.

In effect, 'the EU is following a 'common' rather than a 'single' foreign policy mapped out by bureaucrats in Brussels...EU leaders decide on a common foreign policy, but action is devolved to European capitals'.[11]

In particular this applies to London, Paris and Berlin, whose leaders constituted an informal alliance at the Ghent EU summit in October 2001. Thus the imperative of state interests has taken precedence over efforts to promote institutional initiatives, whether by the UN or the EU, and this is likely in the case of the latter to complicate and delay efforts to forge an integrated foreign and defence policy, with appropriate military capabilities.

And as far as the US is concerned, the Bush administration has preferred to pick and choose European allies on whom it can rely for support measured in both diplomatic and military terms, rather than deal with the EU as a whole. Memories of NATO's time-consuming decision-making over targets during the Kosovo War are still fresh in the US. To this degree, then, US policy remains bilateral in terms of securing helpful military support from key players, while at the same time forging via diplomatic means a global coalition based on the ideological premise of collective security.[12]

Yet even the 'big three' – the UK, France and Germany – are junior partners whose advice and support will no doubt be welcome but not necessarily decisive in the framing of American strategy. Indeed, differences between the allies may well emerge if the US elects to widen its war aims to include, for example, military action against Iraq, long suspected of harbouring and supporting terrorist groups. The treatment of alleged al-Qaeda 'prisoners of war' may well become another such divisive issue.

There is, too, the question of what follows the Taliban's destruction. There is currently a commitment to involve the UN in the rehabilitation of Afghanistan, which would involve the establishment of a broadly based government, massive amounts of aid and a UN presence – both military and strategic – to oversee the work of reconstruction. Creating a functioning nation state on the ruins of a deeply divided and poverty-stricken Afghan society, the structures of which have been ravaged by over two decades of war, will be a formidable undertaking.

Parallels with the international protectorate status of Bosnia and Kosovo come to mind, suggesting yet again that intervention – whether for humanitarian or other reasons – carries enormous costs for the international community, involving an indefinite commitment of human and material resources. What is crucial in any process of nation building, however, is that it must involve drawing on an indigenous Afghan political tradition of thought and behaviour, rather than a wholesale replication of institutions based on Western models.

DOMESTIC IMPLICATIONS

No doubt the events of 11 September inflicted a major dent on the American psyche. This was a compound of horror at what had occurred and a recognition that the US – hitherto seemingly invulnerable to violent attack – was unprepared, both psychologically and strategically, either to deter or defend against such a threat. Moreover the subsequent anthrax scare has raised the spectre of internal disruption by chemical means. How to organise efforts against such threats will become a central preoccupation of all Western governments. The consequences for the maintenance of civil liberties may well be severe: stricter checks on immigration, faster extradition procedures, and intrusions into private modes of communication and financial arrangements could become commonplace.

It is still too early to make any firm prediction about the outcome in terms of the impact on the maintenance of a decent civil society, where the line drawn between the public and the private domains has traditionally been the hallmark of democratic societies.

Moreover, it is significant that the terrorist attacks of 11 September accelerated a downturn in the world economy. In October 2001 alone some 415,000 jobs were lost in the US and the impact on certain industries – tourism and aircraft – has been profound across the West. What is crucial for recovery is the restoration of confidence in business circles, and this may well be more difficult during this recession precisely because the crisis induced by the events of 11 September shows no signs of an early end. Uncertainty about the future has always been a feature of economic activity, but governments and business enterprises have found ways and means of coping with it in the past.

However, the degree of uncertainty generated by the events of 11 September could be said to be qualitatively different insofar as it affects not only policymaking by governments but also permeates day-to-day life. This is suggested by increased security in major airports, checks on postal deliveries, and elaborate precautions to protect the physical symbols of statehood, such as parliamentary assemblies, and military establishments.

We may indeed look back on the decade of the 1990s as the last in which private satisfaction could be pursued without fear of public disaster. Such confidence, some would say arrogance, will be hard to restore.

DISCIPLINARY CONCERNS

Some commentators have argued that the attacks of 11 September are analogous to what followed similarly apocalyptic events in 1789, July 1914 and 1989 in terms of their potential impact on the future of the international system. This is not an unreasonable assumption. We may expect major shifts in scholarly focus and commitment, as old themes and ideas about the structure and process of international relations are either tested for their continuing relevance or jettisoned altogether, in favour of new theories and more appropriate discourses.

Terrorism, for example, will no doubt be subjected to renewed and more rigorous attention, while we may also expect the orthodox premises of contemporary strategic studies to undergo drastic re-examination. How, for example, do we deter and defend against global terrorism? Does, for example, the destruction of the Taliban regime in Afghanistan provide a valuable increment of deterrence against support and safe havens for terrorist organisations in the like-minded states elsewhere? Does the notion of 'virtual war' retain any utility, when real battle involves a willingness to take casualties on the ground in distant, inhospitable places?

Yet another area which will induce even more theoretical speculation is that of the role of the state. Its vulnerability to terrorism has never been doubted, but what – if anything – can be done to restore public confidence in the art of government, and, in particular, to reduce that vulnerability at a time when there appears to be widespread disillusion with the capacity of politicians to handle more conventional domestic issues such as health, education and transport?

Then again there will be a reinvigorated debate about the so-called 'dark' side of globalisation and its impact on Third World states lacking the political and economic capacity of their richer counterparts. Indeed, we may wish to question whether an international society of the kind that Hedley Bull analysed with such skill and insight exist any longer: That society expanded and survived over three centuries via the development of rules, norms, conventions and institutions for regulating and maintaining a degree of international order. Has that society, in both its historical and contemporary manifestations (built as it was on Grotian foundations), been fatally fractured by the events of 11 September?

Or, by contrast, has the crisis generated a belated recognition that rules and norms for good global governance, that can hasten the emergence of a genuine international community of states and peoples, are more essential than ever? Certainly, whether such rules (and the institutions required to legitimise, sustain and enforce them) can be devised is a question that will rightly be debated by thinkers and practitioners alike.

And if such institutions do emerge, they will have to be constructed *ab initio* or based on a Burkean prescription of past and current experience, evolving as time and circumstance dictate. Thus NATO is hardly acceptable as a basis for a world-wide collective security organisation. It carries too much ideological baggage of a kind that would be anathema to states beyond its scope and military jurisdiction. What may emerge is an informal Concert of the Great Powers – not unlike its nineteenth-century predecessor – willing to act quickly, decisively and independently of the UN and its cumbersome decision-making progress.

Such a development would have implications for human rights issues in so far as priority might well be given to ensuring cooperation even if that involved turning a blind eye to abuses within states. And if that is the outcome – and we can only speculate at this very early stage – then the truth of the proposition that 'what goes around comes around' may be emphatically demonstrated. In other words, such a solution would be 'old wine in new bottles', demonstrating yet again that practice – if not high theory – in the realm of international politics shows an astonishing capacity to enter the future backwards, looking to the past to find solutions to unexpected and exceptional future circumstances. Let the debate begin!

REFERENCES

This chapter was prepared in the immediate aftermath of the events of 11 September 2001 and the author hopes that its conclusions remain relevant. He is grateful to the editors - Dr Greg Mills and Ms Elizabeth Sidiropoulos - for permission to reprint the paper following its initial publication in *A New World Order: Implications of 11 September 2001*, published by the South African Institute of International Affairs, Johannesburg, 2002.

1. War is a misnomer in this particular context. Strictly speaking, in terms of international law, war occurs between states, and involves an official declaration and consequent conferment of rights and duties on belligerents. The widespread use of the term by politicians and the media in the present crisis deserves deconstruction beyond the scope of this short chapter.
2. Baylis J. 'Revolutionary warfare' in Baylis J. *et al.* (eds), *Contemporary Strategy* (London: Croom Helm 1987) pp.221–2.
3. Ibid. p.214.
4. Some observers regard this development as a repudiation of fashionable post-modern doctrines of relativism, ironic detachment and rejection of rationalism as a basis for political discourse and progress alike.
5. For an elaboration of this thesis, see Bull H. *The Anarchical Society: A Study of Order in World Politics* (London: Macmillan 1977) and Bull H. and A. Watson *The Expansion of International Society* (London: Clarendon Press 1984).
6. See Simon S. and D. Benjmain, 'America and the new terrorism', *Survival*, Vol.42, No.1 (Spring 2000) pp.59–75; for a detailed analysis of the views of bin-Laden and in particular his belief that US 'aggression' must be countered wherever and whenever it occurs in the Muslim world.
7. I owe this point to Dr John Stone, my colleague in the Department of War Studies at King's College, London.
8. *Financial Times*, 1 November 2001.
9. 'In political activity, then, men sail a boundless and bottomless sea; there is neither harbour for shelter, nor floor for anchorage, neither starting-place nor appointed destination. The enterprise is to keep afloat on an even keel...', Oakeshott M. 'Political education' in *Rationalism in Politics and Other Essays* (London: Methuen 1994) p.127.
10. Bull H. *The Control of the Arms Race* (London: Weidenfeld 1961) p.49.
11. Barber L., 'European 'great powers' break ranks', *Financial Times*, 6 November 2001. This division between 'great' and 'small' European powers was comically illustrated by the spectacle of several European leaders insisting on joining Tony Blair's dinner party for President Chirac and Chancellor Schroder at Downing Street on 10 November 2001. On a more serious note, the Blair government has allegedly proposed a Security Council type structure for the EU in which a triumvirate of the UK, France and Germany would engage in crisis management, etc. This is likely to be strongly resisted at the next Intergovernmental Conference.
12. This notion was suggested to me by Mr Gurdip Nahal, an undergraduate in the Department of War Studies at King's College, London.

3

The War on Terrorism: A New Classic in Groupthink

Cranfield University, Royal Military College of Science

Although most people like to believe that they make rational and logical decisions and expect that national policy-makers would do the same, Simon[1] and many writers since have demonstrated that rationality is limited by human fallibility and/or by environmental constraints. One of the functions of groups is to overcome such limitations but the existence of a group is no guarantee that decision outcomes will be good. Decision-making pathologies do exist and one of these, that can occur when vital decisions are to be made, is groupthink. The flawed decisions that result from groupthink can lead to potentially disastrous outcomes: hence it is important that the phenomenon is well-understood by those charged with making vital decisions within the arenas of defence, security and peacekeeping.

Groupthink has been identified in many defence management situations, including the Bay of Pigs (Cuba 1961) invasion, the escalation of the Korean and Vietnam wars, and inactivity prior to the assault on Pearl Harbor.[2] More recently groupthink has been demonstrated in decisions made by operational commanders during World War II, the Suez crisis and the 1991 Gulf War.[3] It is entirely possible that the response to the events of 11 September 2001 might be the result of such pathology. This chapter offers an assessment of whether the 'war on terrorism' and its associated decisions are the outcomes of groupthink.

A POST-MODERN CONFLICT

Tuesday 11 September 2001 will almost certainly come to be regarded as one of the defining moments in modern history. Everybody can recall when and how they heard the news that New York's World Trade Center (WTC), and the

Pentagon, had been attacked. Through the ensuing days, televisions across the world repeatedly broadcast images of the impacts and the collapse of the towers.[4] The USA is possibly the world's greatest military, cultural and economic power yet, for the first time since Pearl Harbor, it had been attacked on home soil. As that nation, and the world recoiled from the immediate shock, its government could do little else but pledge to bring the perpetrators to justice. President George W. Bush was quick to reassure his people that:

> Our nation, this generation, will lift the dark threat of violence from our people and our future. We will rally the world to this cause by our efforts and by our courage. We will not tire, we will not falter and we will not fail.[5]

From his secure retreat the President denounced the attacks on New York City and Washington DC, and the thwarted hijack attempt in Pennsylvania, as acts of war. He was granted emergency powers to enable a swift response, and brought together a group of his most trusted advisors,[6] many of whom had gained experience during the 1991 Gulf War. Their task was to guide the nation through the months ahead and, importantly, to gather and maintain support from the nation and from the world.

Communities and countries across the globe responded swiftly to denounce the terrorists and to offer condolence and support to the USA in its time of crisis. The NATO allies were the first to answer the call. On 13 September, the NATO Council invoked Article 5 of the North Atlantic Treaty, which declares that:

> an armed attack against one or more [members] in Europe or North America shall be considered an attack against all of them ...[and each member nation will take] such action as it deems necessary, including the use of armed force, to restore and maintain the security of the North Atlantic area.[7]

Individually and collectively, the NATO allies were committed to act against terrorists and those states that offered succour and support to terrorists. The declaration tacitly included NATO's 27 European partners and 7 Mediterranean dialogue countries. Around the world other nations joined in expressing their support for the USA. Among the various visitors to Washington DC in mid-September 2001 were the foreign ministers of Russia and China, the president of Indonesia and several European leaders.[8] Within a few days the White House team, assisted by the United Kingdom government, had built up a coalition of over 130 countries.[9] For those countries that did not offer assistance, George W. Bush gave a very clear message:

Either you are with us or you are with the terrorists[10]

It was intended that war on terrorism was to be 'fought' in numerous ways.

Economic sanctions were introduced to freeze terrorists' financial assets in the G-20 and International Monetary Fund member states,[11] including the USA, and the United Nations passed a new resolution to restate its requirement that national finances should be kept free of terrorist funds.[12] At the same time the US Federal government released $5.1 billion for the 'rescue and reconstruction efforts', including $2.5 billion for the military and $40.86 million for the FBI.[13]

Emergency legislation was passed to make life difficult for known and would-be terrorists. The Proved Appropriate Tools Required to Intercept and Obstruct Terrorism Act (the PATRIOT acronym is, surely, no accident) in the US contained over 50 anti-terrorism bills. These included authority for the indefinite detention of anyone suspected of terrorist connections,[14] and for non-US nationals to be tried by military commissions, rather than civilian courts, and with lower standards of evidence. The United Kingdom enacted the Anti-terrorism, Crime and Security Act 2001[15] which, *inter alia*, also permitted detention without trial for suspected foreign terrorists.

With the assistance of Pakistani diplomats, demands were put before the Taliban authorities, controlling most of Afghanistan, to hand over Osama bin Laden, the prime suspect (with other conditions) or to 'face the consequences'. In the days following 11 September it emerged that some form of attack on Afghanistan was inevitable. The Taliban regime had failed to surrender bin Laden and no-one was entirely certain of his whereabouts.

Operation 'Enduring Freedom' began on 7 October 2001 as, barely one month following the attacks on the World Trade

Center and the Pentagon, US bombers launched attacks on Al Qaeda and Taliban targets and, with support from ground troops, eventually 'liberated the Afghan people from the repressive and violent Taliban regime'.[16] However, as successive British governments have learned from a long experience in dealing with terrorism in Northern Ireland, a struggle against terrorism is not like a war (even the metaphorical 'wars' on drugs or crime). Rather, warns Sir Michael Howard, it is a struggle for the hearts and minds of those who can provide intelligence and deny support to terrorist groups, and bombs are unlikely to achieve that.[17] Even so, within a matter of weeks the *war on terrorism* had become common terminology within the media.

This was to be a strange form of war where nations that were formerly bitter enemies became allies. The coalition included countries, such as Pakistan and Saudi Arabia, that were also suspected of giving succour to terrorist groups, but whose contribution was considered to be vital. Much of the air traffic over Afghanistan's skies was delivering humanitarian aid rather than bombs to show that this was not a fight against the people, but with the ruling regime.

While the USA claimed the justification of homeland defence, according to Article 51 of the United Nations Charter, the target was thousands of miles away from its shores. Further, despite the distance involved, many of those involved could sleep in the safety of their own beds:

> THE B2 Stealth bomber, sleek, black and reeking of science-fiction menace, serves as a kind of metaphor for the disengaged, video-arcade strangeness of modern (*sic*) warfare. Pilots take off from heartland America, sit patiently while they flash across time zones, and deliver 16 2,000lb bombs with computer accuracy …. They then return home…. Never can warfare have been so impersonal.[18]

This was a conflict that overturned normal doctrine and conventions of warfighting, giving cause for concern on many fronts. Strategists note that this was warfare with no declaration of war between states, no clear objective and no exit strategy.[19] Moralists and philosophers argue over the ethics of the conflict, and maybe dwell on the notion of retributive justice. As an organisational analyst, this writer questioned the quality of decision-making that brought the world into this post-modern conflict. Could the dubious emergency legislation being ratified on both sides of the Atlantic, the controversial management of prisoners in 'Camp X-ray', the hastily-formed alliances with unlikely governments, and the air offensive on Afghanistan, be the result of a decision-making pathology?

The first cue came from one of President Bush's many speeches:

> We cannot know every turn this battle will take. Yet we know our cause is just and our ultimate victory is assured. We will, no doubt, face new challenges. But we have our marching orders: My fellow Americans, let's roll.[20]

This self-assured and moralist stance suggested the phenomenon known as groupthink.

GROUPTHINK

The process was described by Janis[21] following his studies into fiascos such as the Bay of Pigs invasion, the escalation of the Vietnam War, and the Watergate scandal. He concluded that strong group cohesion (real or illusory) interfered with critical thinking. Groupthink is a shorthand term to refer to:

> a mode of thinking that people engage in when they are deeply involved in a cohesive in-group, when the members' strivings for unanimity override their motivation to realistically appraise alternative courses of action.[22]

Several antecedents have been identified. The condition is typically likely to occur in high-profile groups, often when they are under stress, and particularly if that stress emanates from some external source. Structural factors such as insularity of the group, homogeneity of members' backgrounds and a lack of impartial leadership are known to increase the likelihood of groupthink.[23]

The outcome is a desire for concurrence that fosters over-optimism, lack of vigilance, and clichéd thinking about out-groups. In many ways, the advantages of group decision making are lost as the consensus-seeking group will, typically shortcut their search for information and limit their assessment of alternative courses of action. Hence, while policy-makers are genuinely seeking the best of possible outcomes from a dire situation, the consequence may be a decision that is not only flawed but, possibly, disastrous.

This is not to imply that all groups (or all governments) suffer from groupthink, although most will display the symptoms from time to time.[24] Neither should we fall into the trap of believing that all decisions that are made with high degrees of consensus are necessarily catastrophic. However, the risks are present: the tragic events of 11 September 2001 put world governments under enormous stress while they attempted to unite in their efforts to respond to, and eliminate, the threat of international terrorism. Moreover, several groups were involved, to varying degrees. President George W. Bush and his advisers (see below) were one group but they formed a sub-set of other groups operating on both a national and an international scale.

Defence management is a sphere that is especially susceptible to groupthink and those involved need to be aware of the potential for such pathology to occur. Given the nature of the current situation and the characteristics of the group(s) involved it should be expected that groupthink will emerge, to some degree, in decisions concerning responses to terrorist threats. It is appropriate, then, that a serious estimation should be made of the extent to which the policy-makers, both within and beyond the USA, will be susceptible to groupthink, and that the quality of subsequent decisions be assessed.

This study was based on a content analysis of contemporary reports and speeches, which were scanned for evidence for or against the major symptoms of groupthink. The aim was to answer the question, can the various ethical dilemmas, legal difficulties and doctrinal problems associated with the offensive on terrorism be explained, if only partly, by groupthink behaviour?

There were two major restrictions on the research.

First, much information relating to the search for bin Laden and the offensive against possible terrorist locations is classified, and likely to remain so for some time. Analysis is, therefore, restricted to reports already in the public domain and the processes of decision making implied from their reported outcomes.

Second, public sources are often the product of some underlying agenda that will be reflected in the content. Hence analysis must consider elisions within given statements, as well as what is actually said or written, in order to apprehend the intentions, underlying meanings and interpretations of both the speaker and the reporter.

Nevertheless, even these limited resources provided sufficient information to expose the symptoms of groupthink. Such evidence is presented here, using Janis' descriptors of groupthink as a heuristic device, taking each of the symptoms in turn.

SYMPTOMS OF GROUPTHINK

Janis identified eight principal symptoms of groupthink, noting that while not all are present in many of the situations he studied, they were seldom observed in non-groupthink decisions. They are grouped into three main types:

Type I   Overestimation of the group, its power and its morality.

Type II   Close-mindedness of group members.

Type III   Pressures toward uniformity.

Type I: Overestimation of the Group

The first grouping of symptoms address the group's estimation of its own abilities. Two particular areas can be identified. Groupthink typically gives rise to decision-making groups holding a high, and possibly unrealistic, assessment of their own power and a belief in their inherent morality. While there might be a dispute over the degree of realism, the Bush administration clearly believed that to defeat Al Qaeda and other terrorist bands was within their capabilities (or those of the coalition). As victims, they could also claim the moral high ground. These points were explicit in a speech from the President, as the attacks on Afghanistan began, and it was this speech that alerted this writer to the possibility of groupthink:

> We cannot know every turn this battle will take. Yet we know our cause is just and our ultimate victory is assured.

We will, no doubt, face new challenges. But we have our marching orders: My fellow Americans, let's roll.[25]

Illusion of invulnerability. With such a huge body of international support it is easy to understand how many people would think that the US could not fail to bring the perpetrators to justice. Unfriendly governments would be unlikely to put up resistance, beyond a token effort, in the face of such power. Thus a perception of military might was established, and supported by remembrance of past achievements: the 1991 Gulf War 'with a decisive liberation of territory and a swift conclusion' and Kosovo 1999 'where no ground troops were used and not a single American was lost in combat'[26]. Such an illusion was further sustained by a belief in the ability of Western technology to keep civilian casualties low.[27]

The coalition also had economic power. Although a great symbol of Western commercial prowess had been destroyed, it did not leave the USA bankrupted. Nor did it undermine a coalition that included the world's greatest economic giants. The USA was able to release massive financial resources for warfighting, homeland security, and to support humanitarian aid to Afghan civilians. Simultaneously, a range of mechanisms were implemented to restrict the flow of funds to terrorist groups.

This illusion of invulnerability also encompassed legislative capacity. Challenges to the emergency legislation in the USA and in Britain suggest that the respective governments misjudged their statutory power. Amnesty International claims that the US PATRIOT Act is, 'too flawed to be fixed'[28] while Irving argues that it flies in the face of International law and natural justice by denying due process to non-US citizens suspected of terrorist activity.[29] In the UK, the emergency legislation has been declared unlawful by the Courts of Appeal as, in allowing the detention of only foreign nationals, it contravened the European Declaration of Human Rights.[30]

Clearly the policy-makers in the US and the UK had an impressive estimation of their power: military, economic and legislative. The terrorists may well have 'roused a mighty giant'[31] but this was a giant that claimed the moral high round.

Belief in inherent morality. Seldom, in human history, has an army marched or a battle begun without the belligerents, severally, claiming that right was on their side. Whether they were Vikings seeking the glory of Valhalla, conquistadors using force of arms to convert native Americans to Christianity, or Japanese kamikaze pilots dying for their Emperor-god, all claimed to be fighting in the name of some greater ideal. Indeed, the Al-Qaeda forces, in a reasoning that seems perverse to others, maintain that they are fighting for the good of Islam and that Allah will give them victory.[32] In the USA in 2001/2002 the theme recurred.

As the victims of atrocity, US citizens understandably wanted to see justice done, while their allies and trading partners in their support for a nation in mourning, desired a similar outcome. The 'declaration of war', however, took the cause further. This was not a campaign simply to apprehend and try those who undertook the attacks on the WTC and the Pentagon:

> This is the world's fight. This is civilisation's fight. This is the fight of all who believe in progress and pluralism, tolerance and freedom.[33]

The White House team was to spearhead an international effort in the defence of Western democracy and security and were, inevitably, convinced of the morality of their cause. This was a utilitarian ethic, based on the morality of intentions, not consequences,[34] so it mattered little that Afghani citizens would be the victims of American bombs, that coalition partners would face moral dilemmas of their own,[35] or that the campaign would polarise the Islamic community and, thus achieve one of bin Laden's goals.[36] But possibilities of moral relativism were dismissed[37] as the morality of the 'Wild West' prevailed: bin Laden was to be captured 'dead or alive'.[38]

Before beginning air raids on Afghanistan, Bush reinforced his moral stance by reminding his people of the atrocities committed by the Taliban against personal freedoms and, in particular of their violence toward women and children. His speeches conveniently omitted reference to the world's long-standing intelligence on these matters; to US reticence in ratifying or complying with international conventions on terrorism; or to the support that the USA had previously given to the Taliban. Clearly, though, the White House team believed in its own morality and, later, the President's supporters would claim that he had fought 'a just war in an honourable, moral way, in keeping with the highest western traditions'.[39]

Given the confidence that the White House, and to some extent the coalition, had in its power and morality, it is understandable that Type II symptoms of groupthink should become apparent. These symptoms related to the ways in which the consensus-seeking group deal with information.

Type II: Closed Mindedness

Pressures for unanimity, within a decision-making group, induce a restricted ability for group members to handle information and to free themselves from dogma and clichés in their thinking.

Collective effort to rationalise and to discount warnings or other information. From the moment when George W. Bush called his nation, and the world, to unite against terrorism it has seemed that some form of military offensive would be inevitable. While military planners have advised of the folly of a direct offensive on Afghanistan, White House strategists failed to find a better plan. Without inside knowledge of discussions within the White House, or of conversations with other coalition leaders, and lacking access to intelligence reports, it is impossible to judge the extent to which such advice is being heeded.

The CIA admitted that 'Disabling the Afghan airforce is about as impressive an achievement as sinking the Swiss navy',[40] and Northern Alliance chiefs pointed out that Taliban fighters were not in their training grounds and that the airfields contained a grand total of only 12 aircraft.[41] Western reporters were advised that any violent response would be playing into bin Laden's hands.[42] Sending ground troops into such a hostile terrain, especially as winter approached, was clearly a hazardous enterprise. Yet, still, the bombers flew, warships set sail for the Indian Ocean and armies were deployed to the region.

Alternative views and proposals do exist but they are not officially noted and discussion of such matters has been relegated to the websites of anarchist, anti-capitalist, pacifist and conspiracy theory groups.[43]

Meanwhile, there is evidence that the USA's response is subject to rationalisation.

At the time of writing, the Taliban has been all but ousted from Afghanistan but there is no clear knowledge of the fate, plight or whereabouts of the prime suspect. Over a few months the objective and priorities of the military attack were re-framed. The purpose of the offensive on Afghanistan evolved so that the objective was apparently not to capture bin Laden 'dead or alive' (see above), but to overthrow the Taliban and install a stable, democratic government in Afghanistan.[44]

A new government has been established in Afghanistan but as the result of intervention by the international community. This was not the result of a democratic process but of arrangement made by the coalition powers, and this 'government' includes elements of the Northern Alliance (which had previously been condemned as a terrorist organisation). On-going international efforts: sanctions against terrorism and the Taliban;[45] humanitarian aid to the citizens of Afghanistan; and mine clearance have been directly attributed to US and coalition efforts.[46]

Furthermore, many old enemies, such as Iraq and North Korea and Sudan were named as possible targets[47] for coalition efforts not just because they may support terrorism, but because they comprised an 'Axis of Evil' that threatens the security of the West. While the intention might not be wholly disingenuous, justification was claimed from Article 51 of the UN charter[48] which allows a member state to act in defence of its homeland.

While there are few examples available to demonstrate collective rationalisation, further evidence of narrow-mindedness is clear in the way that the 'enemy' was stereotyped.

Stereotyped view of enemy leaders as evil. In the aftermath of 11 September 2001, as coalition leaders rushed to condemn the atrocities, they were vociferous in their reprobation of the enemy. In one speech alone President Bush described the enemy (whether that was the terrorists or the Taliban is unclear) as extremist, perverted, brutal, unreasonable, repressive, murderous, evil and treacherous:

> We have seen their kind before. They're the heirs of all the murderous ideologies of the 20th century. By

sacrificing human life to serve their radical visions, by abandoning every value except the will to power, they follow in the path of fascism, Nazism and totalitarianism. And they will follow that path all the way to where it ends in history's unmarked grave of discarded lies.[49]

Bin Laden came in for special castigation, his evil made 'Saddam Hussein look like Jimmy Stewart'[50]. Such a clear stereotype overlooks the, albeit malevolent, genius that could mastermind such a complex operation as an assault on mainland USA. Additionally, it takes no account of the strength of the al-Qaeda network (and its supporters), which could have as many as several thousand members and serves as an umbrella organisation for Sunni extremists such as the Egyptian Islamic Jihad and the Mujahadin.[51]

Further, this was a picture of people so evil as not to warrant serious regard, or basic human rights:

> Those who practise terrorism – murdering or victimizing innocent civilians – lose any right to have their cause understood by decent people and lawful nations.[52]

This invites serious problems for fighting troops. Hundreds of people were detained without trial at Camp X-ray in the US naval base of Guantanamo Bay, Cuba, having been categorised as neither criminals nor prisoners of war. If the US, and other coalition powers, were prepared to disregard the Geneva Conventions in respect of their prisoners they could hardly expect other than equal treatment for their own nationals if they should be captured.

Nevertheless, this stereotype that was shared by others beyond America's shores. The moral stance, set by the White House, the international consensus on the need to 'do something', and a desire to show solidarity may have been factors that obliged individuals and nations to acquiesce to US plans. Pressures that coerce conformity form the third set of symptoms that Janis identified with groupthink.

Type III Pressures toward Conformity

Building up a consensus on several levels: within the policy-making team; within the USA; and within the wider, international community was essential to establishing and maintaining support for the offensive on international terrorists:

> all of the people who have been speaking to the American people have been talking about the need to capture the momentum that has built up after the 11th of September into building a coalition against terrorism. And that is something for us, anyway, that seems to be going extremely well. (Marc Grossman, Under-Secretary of State for Political Affairs)[53]

In any group that is seeking concurrence between its members, notes Janis,[54] there is very little scepticism over the clichés, stereotypes and assumptions that frame decisions. When objections are not voiced, the impression that decisions are rational and viable is reinforced. However, the objective of unanimity leads to various subtle or explicit strategies to silence dissenters. Janis identifies four forms that such strategies can take.

Self-censorship of deviations from consensus. When, in an atmosphere of mutual respect, a group of people arrive at a unanimous opinion, each member of the group is liable to feel that that view must be correct. This 'consensual validation'[55] begins to replace individuals' critical thinking unless there is open discussion and disagreement among group members.

We cannot know how many people disagreed with the White House decision to launch the bombardment and invasion of Afghanistan because it is in the nature of groups to develop and maintain consensus about group decisions and behaviours. Indeed, it has become an accepted certitude of social psychology that individuals within groups experience clear discomfort if their own views deviate from the group norm, and that individuals will tend to maintain silence rather than reveal their own doubts. This phenomenon increases with the apparent unanimity of the group.[56]

As a consequence there is little clear evidence of the extent to which individuals are censoring themselves. However, for Republican politicians, to deviate from current opinion would probably be a matter for resignation and the end to a political career. Democrats, meanwhile, cast in the role of a loyal opposition at a time of national crisis, have also felt the need to keep silent rather than deviate from the national norm.[57]

Meanwhile, across the Atlantic, there appeared to be whole-hearted support for the offensive from the British government and opposition. However, it has later been reported[58] that the British Foreign Secretary had urged constraint, voicing departmental concerns that any attacks on Afghanistan could have violent repercussions throughout the Middle East. As a senior member of the government, though, he was obliged to stand with the Prime Minister when it was announced that British troops were to be deployed. His body language, described as 'wretched',[59] betrayed his discomfort while his presence apparently declared his support for the Prime Minister's speech.

It follows that, where objections are not raised and concerns are not shared, the illusion of consensus is strengthened. The sharing of such an illusion is next on Janis's list of symptoms.

Shared illusion of unanimity. In a consensus-seeking group there is a tendency for group members to support each other, and to emphasise the points of agreement at the expense of permitting free and open discussion. The result is that the group *appears* to be in agreement, although that agreement may be illusory.

NATO's invocation of Article V of the North Atlantic Treaty and the flock of visitors to Washington DC during September and October 2001[60] gave an impression of widespread international support. While support is not synonymous with agreement, George Bush was willing to declare that 'The civilised world is rallying to America's side.'[61]

This apparent unanimity, at the international level, disguised the depth of debate and disquiet within those countries, individually or severally. The British Foreign office (above) was concerned about the consequences of military action. Anderson told how Turkey was worried about the reaction of Muslim nationals but, as a NATO member, was bound by the decision on collective defence; and the same report tells that 'Pakistan had misgivings ... the same went for India.'[62] Other countries with a substantial Muslim community faced similar difficulties.

At the American national level, growing numbers of anti-war websites and rallies give the lie to the impression that the whole country has been behind the campaign against terrorism. If, as the right-wing press claimed, President Bush had a 90 per cent approval rating,[63] then ten per cent of the population did not support the president's response to the terrorists.

Within President Bush's select inner group of advisors Secretary of State, Colin Powell, is known to have advised caution and was criticised for being too reluctant to use force.[64] Possibly as a result, he was allocated to the role of shuttle diplomat so that he was on successive overseas tours while the White House group was making its, apparently consensual, decisions.

Thus the impression that the policy-making group, the nation, and the world were solid in support of the *war on terrorism* was maintained as an illusion. Where there are dissenters, a team taken up by groupthink will find ways of putting them under pressure to conform – Janis's next symptom.

Direct pressure on dissenters. Part of the reason that unanimity can be illusory is that those who might dissent or oppose the chosen course of action feel that they cannot give voice to their opinions. Self-censorship (above) occurs when people believe that they must be the only dissenter(s) because others appear to be in agreement. More ominously, the consensual group might bring direct pressure to bear on the dissenters, to oblige them to maintain the illusion of unanimity.

On the international scene the USA, as a leading military and economic power and a major donor of aid, 'can exercise a salutary influence'[65] upon poorer nations. Bearing that in mind, it can be appreciated that Bush's *Address to the Nation*[66] gave a clear, if implicit, message to countries with otherwise tenuous relationships with Washington. Either they should support an offensive against terrorism or face US retaliation.[67]

Pakistan, for instance, has held a key role in the assault on Afghanistan. It was one of only three countries that recognised the Taliban as a legitimate government, but it was essential that it should allow the use of its airspace for coalition forces and that the borders should be closed to contain al-Qaeda forces. General Musharraf, the president of that Muslim country was put under 'the most ferocious pressure'.[68] He could not afford to make an outright enemy of the USA so he found himself:

> caught between a rock and a hard place. On the one hand, the Bush administration is saying ... you are either with us or against us – choose which side you're on. On the other, hard line religious opinion ... will react angrily if the government helps American forces.[69]

Within the USA those who believed that military intervention was not the best course of action risked being labelled as unpatriotic.[70] When Democrat Senate Majority Leader, Tom Daschle, expressed concern that the conflict was expanding without a clear direction,[71] he was publicly rebuked by fellow senators:

> How dare Senator Daschle criticise President Bush while we are fighting a war on terrorism in the field. He should not be trying to divide our country while we are united' (Senate Minority Leader, Trent Lott)[72]

There is also evidence to suggest that pressure was also being brought to bear on members of Bush's advisory group. For instance, Attorney General John Ashcroft, has a complex and vital role that includes taking the brunt of any criticism for the team. His job has been likened to that of a lightening rod but he has been put under pressure to demonstrate that his deeply-held Christian views will not 'affect his job'.[73]

A more subtle form of maintaining consensus within a decision-making group, is to ensure that contradictory information does not affect the resolve of the group leader. This role falls to 'mindguards' who may or may not be self-appointed.

Mindguards. A mindguard is someone who protects the decision-making group from adverse information that might undermine the group's perceptions of the morality and effectiveness of its plans. Mindguards could be likened to bodyguards where they are protecting ideas rather than persons. Remembering that not everyone in a group need be a mindguard, within the current context it is useful to consider who is advising and informing the President of the USA.

Operation Enduring Freedom was the brainchild of a small group of White House advisors that had formed a de facto war cabinet to advise and inform a president who came to power with no foreign policy experience. These people were selected for their posts, not solely because of relevant expertise but also, for their loyalty to party-political ideology. Many have links within the oil industry, which has a vested interest in establishing a stable government in Afghanistan. Moreover, all but one of this select group occupied similar positions in the administration of George Bush Senior, supporting and advising him through the 1991 Gulf War. For many of this team, the Gulf War was never concluded and the current conflict presents a means of completing the unfinished business of 1991.[74]

- Vice President Dick Cheney has been described as a father-figure to the current president and as 'A true Washington insider'.[75] He was Secretary of Defense during the 1991 Gulf War.
- The current defense secretary, Donald Rumsfeld, is recognised as, one of the White House's leading 'hawks'. He is known to hold a hardline on defence issues and was tasked with directing the USA's military response to the threat from al-Qaeda.
- Rumsfeld's deputy, Paul Wolfowitz, shared the enthusiasm for a 'bold and sustained war on terrorism'[76] and was described as 'emerging as the main strategic planner behind a possible assault, which he has said could aim so far as to "end" states which sponsor terrorism'.
- Secretary of State Colin Powell had been expected to be a star of the current administration. However, his Clausewitzian approach to defence was not popular. It has been speculated that the President actually takes foreign policy advice from his National Security Advisor Condoleeza Rice.[77]
- Dr Rice is said to be very close to the President, who credits her for tutoring him on foreign affairs. She is known to be enthusiastic about the missile defence shield and believes that the USA has no choice but to carry on the conflict into Iraq.[78]
- Presidential counsellor, Karen Hughes, is reported to be able to explain complex issues in 'simple terms that Bush can

understand'.[79] Apparently, on the eve of his inauguration Bush told his staff 'I don't want any important decisions made unless she's in the room.'[80] *The Times* (of London) reports that 'She [Hughes] has always been very useful for other aides to bounce ideas off first ... She would say things like "Don't do that, he's in a bad mood today".'[81]

It must be remembered that vast amounts of information pass through the White House, and that the President needs to trust his staff to decide what warrants his attention. During the state of emergency, and the ensuing offensive against terrorism, these were the people tasked with advising the President over strategy, and with filtering the vast amounts of information that are put before him.

Thus there was, to varying degrees evidence of all the symptoms and antecedents of groupthink that Janis has identified and the factors were in place that could lead to fiasco in decision making. The international coalition was committed to military action against some of the poorer countries of the world based on decisions made by a cohesive group within the White House. This was a group that holds a high estimation of its power and its morality; that has exhibited close-mindedness; and was capable of exerting strong pressure to maintain uniformity of opinion. Indeed, the group influence spread beyond the Oval Office, into the nation and much of the rest of the world.

Outcomes of Groupthink

One of the curiosities of the attack on Afghanistan was that, at the same time that bombs were being dropped on Taliban targets (and, inadvertently, in other areas), humanitarian aid was being sent into the country by a variety of means and much of it from the USA. This demonstrates one of the paradoxes of groupthink. The decision-making group in Washington was not made up of nasty people but of normal human beings who held ethical principles. They had families and friends and were all well-respected for their expertise. However, in the groupthink situation, resulting from an atmosphere of consensus, people find it relatively easy to take a hard line and to authorise actions that would otherwise be difficult and controversial. The tendency to seek unanimity fosters a lack of vigilance, clichéd thinking about the 'other' and prevents individuals from raising those moral questions which could imply that:

> …this fine group of ours, with its humanitarian and high-minded principles, might be capable of adopting a course of action that is … immoral.[82]

The outcome of groupthink is poor quality decision-making. It could be argued that the bounded rationality which limits individual decision making, and which groups are supposed to overcome, is liable to re-emerge in consensus-seeking groups. It is useful, at this point in the discussion, to consider the deficiencies that Janis has identified. Although it is difficult for anyone outside President Bush's advisory group to assess the extent to which decision making was limited, several of Janis's exemplars have parallels in the current context.

Within groupthink situations discussions are usually restricted to considering a limited set (Janis suggests that this might be as few as two) of alternative courses of action. After 11 September 2001, there was a brief diplomatic effort and economic sanctions were introduced. However, the swift, bellicose response to the attacks limited the scope for diplomatic intervention. It further prevented those who planned or abetted the attacks being declared criminals, to be pursued by international law. The aerial attacks on Afghanistan were unlikely to win the 'hearts and minds' of those that could deny support to terrorist groups.[83]

As a result of groupthink, the setting of objectives is liable to be incomplete, with decision making groups failing to consider the values that are implicated or compromised by the choice. The 'war on terrorism' is a war with no geopolitical boundaries, no limit on who may be an enemy and no limit on weaponry. President Bush's pledge on the night of 11 September was to find and deliver justice to those who committed such atrocity. That was quickly extended to include those governments and peoples that provided succour to terrorist groups. All too soon an 'Axis of Evil' was identified, to be included in the category of enemy and by mid-2002 the world was waiting to see if the USA was prepared to make pre-emptive attacks against Iraq. All this in an environment where there was no acceptable definition of terrorist or terrorism.

A groupthink decision typically involves a failure to properly evaluate the preferred choice with regard to risks and the possibility of failure. It must be assumed that this has happened within the White House group, for over a year following the attacks on the USA, bin Laden had not been apprehended[84] although many people were being detained without trial in Cuba with the novel and dubious classification of 'unlawful combatant'. Former enemies (Russia) and a group previously regarded as terrorists (the Northern Alliance) were controlling parts of Afghanistan. The delicate relations between India and her neighbours has been upset as, for instance, India declared the guerrilla fighters of Pakistani Kashmir to be terrorists and accused the Pakistan government of supporting them.

Alongside these failings it may be expected that the decision- makers had spent little time discussing previously-rejected alternative choices (assuming that alternatives were considered) to assess whether they might offer advantages that had been overlooked. This would be compounded by a poor search for information with little or no attempt to gather the opinions of experts from beyond the group. Rather, what occurs within the consensus-seeking group is a selective bias in the way factual information and opinions are processed.

Evidence for such bias comes from the response to the anthrax scare. In a heightened state of alert, the US government was warned by its intelligence agencies that terrorist groups might try to attack with biological weapons so when anthrax spores were found in various mailed packages it was initially assumed that it came from a (foreign) terrorist source. Valuable time and effort was lost in tracing the source.

While no social scientist should ever claim that their case is proven it is appropriate to suggest that the available evidence supports a particular stance. This chapter began by asking whether the ethical dilemmas, legal difficulties and doctrinal problems associated with the offensive on terrorism could be explained, if only partly, by groupthink behaviour. It has subsequently shown that the antecedents of groupthink were present. It has offered evidence of the individual symptoms of groupthink and, to a lesser extent, of the likely outcomes in terms of the decision-making process. The short answer to the question, then, is 'Yes'. It may be that other decision-making pathologies are also present in some degree but that is beyond the scope of this discussion.

It is depressing to realise that, 30 years after Janis' original research,[85] the administration of the one of the world's great powers is capable of succumbing to the perils of groupthink. This is not because they are unaware of the phenomenon, for the USA is a well educated and self-aware society, but there is denial to the point where some of the more extremist press are accusing the Islamic world of being locked into the groupthink process.[86] If denial continues the outcome may yet be another fiasco, of the magnitude of those described by Janis, with grave results for global peace, security and stability. There is however a range of tactics that, while uncomfortable, may be employed to arrest the process.

AVOIDING GROUPTHINK

The first step to reducing the effects of groupthink is for any decision-making group to realise and accept that it may be susceptible. The responsibility lies with each group member but particularly with the leader.

Impartial leadership is the key to avoiding groupthink. For the leader of a cohesive group to express personal views or preferences ahead of a debate limits the extent to which individuals will be willing to expose their own doubts or preferred options. Rather, group leadership should encourage discussion, the search for clarification and the development of alternatives. An impartial leader allows group members time to consider a decision before it is implemented. Individuals might, for instance, wish to consult with their own advisers or check particular facts or suppositions with trusted experts.

If this is to work, though, leaders need to be genuinely non-directive and prepared to cope with their personal frustrations if the group arrive at an option that is contrary to the leader's personal view. Within an environment that is defined by party politics and the need to retain popular approval, this is problematic. Surrounded by party faithfuls and, particularly, by people who are known to be in favour of a martial response, it will be difficult for President Bush to maintain an impartial stance.

A group leader needs to become the critical evaluator for each of the team members. Individual group members ought to be able to air their objections, doubts and other opposing opinions. If the group leaders show a willingness to accept criticism, or refining, of their own ideas, and a disposition to be influenced by the contributions and suggestions of others, that helps to create an atmosphere where dissenters can disclose their doubts and shatter illusions of unanimity. Sometimes it is consensus rather than disagreement that must be managed.[87]

It has become a fairly common practice that policy-making groups appoint one of their members to the role of Devil's advocate but that it not necessarily a fail-safe strategy for, where consensus is high the Devil's advocate can still be taken up by the groupthink influence so that objections are kept deliberately mundane or self-defeating. The purpose of such a role is the avoidance of tokenism, so at policy-making meetings sizeable chunks of time should be devoted to surveying warning signals.

For this reason Janis[88] suggests the additional appointment of a 'Cassandra's advocate' to focus attention on the most alarming of possible consequences of the chosen course of action. This is particularly relevant where groups are making decisions that involve relations with other nations.

If the group deliberately sets about constructing different scenarios, it is possible that some new insights can be gained in interpreting antagonistic behaviour. For instance, by asking 'what did al-Qaeda gain by attacking the USA?', it is possible that the group might have recognised it was an act of provocation rather than of war, leading to an entirely different response from the White House.

Another prophylactic against groupthink is to solicit and listen to other opinions. Experts in particular aspects of the issue(s) under consideration can be brought into the group meetings, on a staggered basis to challenge risky views. This was a strategy that President Kennedy used at the time of the Cuban missile crisis and which helped to prevent a repetition of the poor decision-making that had resulted in the Bay of Pigs invasion.[89]

Remarkably, the proposal to launch an offensive against Iraq may be the factor that will break down the groupthink behaviour. During 2002, White House policy-makers faced a rise in adverse domestic[90] and international opinion.[91] The movement away from consensus began with various world leaders, distanced from the consensus-seeking group, making their concerns known. During the time of writing this chapter, for instance, King Abdullah of Jordan has made his feelings known:

> All of us are saying 'Hey, United States, we don't think this is a very good idea.'[92]

Other Middle Eastern states (Egypt, Syria, Iran, Lebanon) have stated that they would not be able to support a pre-emptive strike.[93]

As this book went to print the debate was continuing. Voices were being raised in objection to an attack on Iraq from people who are at least as equally expert in international relations, politics or warfighting as the White House group. They came from diverse cultures and all had various agendas in dealing with terrorism. As a result of groupthink, it is entirely possible that President Bush's advisory group would have difficulty accepting other points of view but, as international support recedes, opposing perspectives are bound to become more prominent.

For instance, Tony Blair came under criticism from the Opposition and from many of his own party for committing UK forces to the 'war on terrorism' without allowing proper, parliamentary debate.[94] Physical distance separated Mr Blair from the rest of the policy-making group. Without the emergency powers that have been granted to President Bush, he was obliged to work with and not against Parliament. If Britain and the European Union should reduce the degree of support that is being offered to the USA, the White House group may be obliged to rethink its strategy.

The 'war on terrorism', as it has unfolded so far, shows the features of groupthink and consequent defects in decision making. The result has been a hasty decision to take a military offensive on an impoverished and oppressed nation (although Bush stated that the war was not with the Afghan people), based on dubious ethical, legal and doctrinal principles. The outcome may yet be a (expensive) fiasco with grave results for global peace and security. However, there is still time to arrest the process and to prevent further damage to international relations. The lessons that can be learned are relevant to anyone in the defence management environment. Principally the members of any policy-setting group need to recognise that they are susceptible to groupthink and to make a clear commitment to high quality decision making.

REFERENCES

1. Simon, Herbert, 'Rational Choice and the Structure of the Environment', *Psychological Review* No.63, pp. 129–38 (1956); see also York's chapter in this volume
2. Janis, Irving *Groupthink: Psychological Studies of Policy Decisions and Fiascoes* (Boston: Houghton Mifflin 1982).
3. Bordin, Jeffrey, 'On the psychology of moral cognition and resistance to erroneous authoritative and groupthink demands during a military intelligence analysis gaming exercise', Joint [US] services conference on professional ethics, Springfield, VA, January 2002.
4. Jaworski, Adam, 'Busy saying nothing at all', *The Times Higher Education Supplement*, 21 September 2001.
5. *Washington Post* 'President Bush Addresses the Nation' <www. washingtonpost.com/wp-srv/nation/specials/attacked/transcripts/ bushaddress_0>, 20 September 2001.
6. Straub, Bill 'Bush's 'war cabinet' seasoned: Aides for president with no foreign affairs experience', *Detroit News*, 27 September 2001; 'Bush's war cabinet' *Straits Times Interactive* <www.straitstimes.asia1.com.sg/mnt/ webspecial/WTC/wtcnews.html> accessed, 11 July 2002; Branson, Louise, 'President George W. Bush's 'war' cabinet' *Straits Times Interactive*, <www. straitstimes.asia1.com.sg/mnt/webspecial/WTC/war cabinet.html> accessed, 11 July 2002.
7. NATO/OTAN *The North Atlantic Treaty* <www.nato.int/docu/basictxt/treaty. htm>, 1949.
8. BBC News, 'US finds allies in anti-terrorism war', *BBC News*, <www. news.bbc.co.uk/hi/english/world/europe/newsid_1542000/1542154.stm>, 13 September 2001.
9. Coalition information centres 'The global war on terrorism: The first 100 days', White House press release <www.whitehouse.gov/news/releases/ 2001/12/100dayreport.html>, 2001.
10. *Washington Post* (note 5), 20 September 2001.
11. Coalition information centres (note 9).
12. United Nations Security Council Resolution 1267, 15 October 1999.
13. The White House *News release*, <www.whitehouse.gov/news/releases/2001/ 09/20010921-8.html>, 14 September 2001.
14. Irving, Molly, 'Patriot games', *Creators syndicate* <www.workingforchange. com>, 31 October 2001.
15. BBC News, 'Anti-terror Act at a glance', *BBC News* <news.bbc.co.uk/1/hi/uk_ politics/170435.stm>, 14 December, 2001.
16. Coalition information centres (note 9).
17. Howard, Michael, 'It was a terrible error to declare war', Lecture to Royal United Services Institute, London reported by *The Independent*, 2 November 2001.
18. Christopher Anderson, 'It's Bush's war – but Blair is the brains', *Mail on*

Sunday, 14 October 2001.

19. See Jack Spence's contribution in this volume.
20. Bush, George W. speech at World Congress Centre, Atlanta, Georgia, 8 Nov. 2001 quoted at <www.thejustcause.com>.
21. Janis (note 2).
22. Ibid.
23. Ibid.
24. Ibid.
25. Bush (note 20).
26. *Washington Post* (note 5).
27. Holmes, Kim 'Maxims for conduction a war on terrorism', *The Heritage Foundation* <www.heritage.organisation/shorts/20011105.html>, 5 November 2001.
28. Amnesty International 'Us Military Commissions: Second class justice', <www.Amnesty.com/AMR51/049/2002.html>, 22 March 2002.
29. Irving (note 14).
30. 'Court rules terror suspects' detention unlawful', *The Guardian*, 30 July 2002; The British government's appeal against this decision is due to be heard on 7 October 2002.
31. The White House Press Release <www.whitehouse.gov/news/releases/2001/09/20010916-2.html> 16 September 2001.
32. 'Aljazzera' *New York Times magazine*, 18 November 2001.
33. *Washington Post* (note 5).
34. Marcus, A. A. *Business and Society: Ethics, Government and the World Economy* (Boston, MA: Urwin 1993).
35. Lindlaw, Scott, 'Bush leans on Pakistan's president', Associated Press, 29 December 2001.
36. 'US leaders warned against war', BBC News Wales <news.bbc.co.uk/hi/english/uk/wales/newsid_1547000/1547674.stm>,17 September 2001.
37. Guiliani, Rudolph speech to the United Nations quoted at <www.thejustcause.com>, accessed 31 July 2002.
38. The White House, News release <www.whitehouse.gov/news/releases/2001/09/20010917-3.html>, 17 September 2001.
39. Buchanan, Patrick J., 'George W. Bush, Man of the Year, 2001: War Leader in highest western tradition' *Human Events Online.* <www.humaneventsonline.com/articles/12-17-01/buchanan.html>, 17 December 2001.
40. Anderson (note 18).
41. Ibid.
42. BBC Wales 'US leaders warned against war', *BBC News Wales*, <http://news.bbc.co.uk/hi/english/uk/wales/newsid_1547000/1547674.stm>, 17 September 2001.
43. See for example: <www.democraticunderground.com; globalrcscarch.ca>; <www.fromthewilderness.com>.
44. Brownfield, Allan C., 'Questions grow over war on terrorism', *Jane's Terrorism and Security Monitor*, http://www4.janes.com, 1 April 2002.
45. United Nations Security Council Resolution 1267 (note 12).
46. Coalition information centres (note 9).
47. BBC News 'America widens 'crusade' on terror', *BBC News*, <http://news.bbc.co.uk/hi/english/world/americas/newsid_1547000/1547561.stm>, 16 September 2001.
48. United Nations Charter, *Article 51*, 24 October 1945.
49. *Washington Post* (note 5).
50. Buchanan (note 39).
51. United States Department of State, 'Al-Qaeda', *Patterns of Global Terrorism, 2000*, <www.library.npr.navy.mil/home/tgp/Qaeda.htm>, April 2001.
52. Guiliani (note 37).

53. Straub (note 6).
54. Janis (note 2).
55. Ibid. p.37.
56. Asch, S., 'Studies of independence and conformity: a minority of one against the unanimous majority', *Psychological Monographs* Vol. 70 (1956) pp.116.
57. Bernard Weiner 'War on terrorism for Dummies' *Democracy means you*, <http:// democracymeansyou.com/serious/dummies.htm> accessed 9 July 2002.
58. Anderson (note 18).
59. Ibid.
60. BBC News (note 8).
61. *Washington Post* (note 5).
62. Anderson (note 18).
63. Buchanan (note 39).
64. Straub (note 6).
65. Mason, Barnaby, 'Analysis: building a coalition', *BBC News*, <http:// news.bbc.co.uk/hi/english/world/americas/newsid_1546000/1546289.stm>, 16 September 2001.
66. The White House, 'President's Address to the Nation', <www.whitehouse. gov/news/releases/2001/10/20011008-3.html>, 8 October 2001.
67. Newton, Christopher, 'US moves to shore up support for war', *Associated Press*, 19 December 2001.
68. Mason (note 65).
69. Lindlaw (note 35).
70. 'Unpatriotic? No, dissidents are as American as can be, *San José Mercury News*, Editorial 1 October 2001.
71. Meyer, Dick, 'Tom Daschle: Potemkin villain', *CBS News*, <www.cbsnews. com/stories/2002/03/07/opinion/main503212.shtml>, 7 March 2002.
72. Brownfield (note 44).
73. Branson (note 6).
74. Bonney, Norman, 'Gulf War II: like father, like son', *The Times Higher Education Supplement*, 21 September 2001.
75. *Straits Times* (note 6).
76. Straub (note 6).
77. *Straits Times* (note 6).
78. BBC News, *Breakfast News*, 15 August 2002.
79. Allen-Mill, Tony, 'Bush exposed as guardian angel flies off', *Sunday Times World News*, 16 June 2002, p.26.
80. Straub (note 6).
81. Ibid.
82. Janis (note 2) p.12.
83. Howard (note 17).
84. See Jack Spence's chapter in this volume for a discussion of the likely dilemmas that would arise from the capture of bin Laden.
85. Janis (note 2).
86. West, Diana, 'Islam's groupthink', *Washington Times*, <www.washtimes.com/ op-ed/20020308-772501.htm>, 8 March 2002.
87. Harvey, Jerry R., 'The Abilene paradox: the management of agreement', *Organizational Dynamics* Vol.17, No.1 (1988) pp.17–43.
88. Janis (note 2).
98. Ibid.
90. 'US debates Iraq war fears', *BBC News*, <http://news.bbc.co.uk/1/hiworld/ Middle East/2163183.stm>, 31 July 2002.
91. Brownfield (note 44).

92. 'Jordan urges restraint over Iraq', *BBC News*, <http://news.bbc.co.uk/hi/english/world/Middle East/2158081.stm>, 29 July 2002.
93. BBC News, *10 O'clock News*, 30 August 2002.
94. Newman, Cathy, 'Most of Cabinet "may oppose war on Iraq"', *Financial Times*, 2 September 2002.

4

Jointery: Military Integration

TREVOR TAYLOR

Cranfield University, Royal Military College of Science

This writer believes that 'jointery' will be of increasing interest to those governments that seek to optimise the effectiveness of their armed forces and the defence capabilities that they can generate from the funds that they allocate to defence.

Definitions are significant here, not least because jointery does not appear yet in dictionaries of English. Jointery is an umbrella term covering activities that either coordinate or integrate the activities of individual branches of the armed forces of a state. Such activities can cover the operation of military forces, the preparation of military forces (most obviously training) and, closely related, the management of the resources used by the forces.[1]

However, as will be argued in more detail towards the end of the chapter, jointery can also address activities of the armed forces and other organs of the state.[2] Even some non-governmental actors may become involved.

This chapter will emphasise that the pressures for more jointery come from two main directions: the perceived need to effective military performance in an age of rapid technological change when the armed forces of major states are regularly engaged in operations; and the pressures to spend every defence dollar wisely in an era when resources for defence are scarce in many states. As these pressures continue to operate over time, jointery will not reach a steady state – indeed it should be seen a 'process' (an almost indefinite series of changes) rather than a condition that some armed forces will achieve.

In this jointery has much in common with European integration: in particular the concept of 'spillover' is relevant to both spheres. Spillover involves cooperative activity in one area tending to deepen and widen to related areas over time so that all the benefits sought from the original cooperation can be captured.[3]

However, it must be recognised that jointery can only be successful in the context of a defence policy and strategy that defines the security challenges to a state and articulates the place that armed forces are to play in their control and resolution. Coherence of defence effort is possible only within the framework of a coherent and specific policy that provides direction to those shaping armed forces. In the UK, two key documents are the *Strategic Defence Review* of 1998, setting out defence policy, and *British Defence Doctrine*,[4] setting out a defence-wide conceptual and practical base for the preparation of military forces to support that policy.

A final opening observation is that this chapter makes frequent reference to the logic pursued in the United Kingdom. As will be noted at the end, this logic is finding some resonance in other European states but it must be stressed at the outset that the United States is different. The contrast between developments in the UK and the US can be ascribed to two factors. First and obviously, US forces operate on a much larger scale than those of any European state. Many US single service support activities operate therefore on an economic scale and there would be little to gain by merging them into a joint body.

Second, the US stresses from its constitution downwards that power should not be too much concentrated in one institution and, to a degree, competition among the services is enjoyed rather than tolerated. Congress can play one service off against another. The organisation of armed forces on an extensive joint basis is only for governments that have confidence that their armed forces, if integrated closely at the top, will not have a worrying amount of power.

Let me begin with an outline of how the UK used to organise its armed forces and many countries still do.

Traditional thinking is that armed forces have separate branches that are expert in operating in a specific environment. Each branch has its own budget which it allocates as it sees fit to personnel, training, equipment, infrastructure, support and so on. Each branch also tends to compete against the others for resources allocated to defence.

The Minister of Defence, most obviously by specifying top level guidance through defence policy and strategy, seeks to impose some coherence on the three branches and stands in judgement on their competing claims. One route to a quiet life for the ministry is to establish accepted percentage shares of the defence budget for each service. As the budget changes, each service keeps its share of the total.

The essential advantage of such a system is that each branch has control of all its activities and can seek to optimise how its resources are used so as to maximise its contribution to overall capability. The cardinal disadvantages are the limited coordination of defence spending that can be forthcoming; and the needless duplication of effort that can be involved, especially in support areas. A lack of interoperability in communication systems is also common. The capacity of separate branches easily to undertake joint operations is usually limited.

Reinforcing any view that the services should be separated for most purposes is recognition that they usually have different cultures that particularly reflect their relationship to equipment. Even in the developed states of the twenty-first century, armies are collections of people who are supported by a wide range of equipment. The fighting elements of navies, on the other hand, comprise large capital assets which are operated and fought by tens, hundreds, or even thousands of people who form an interdependent community. All people in a ship must work together to make her operate successfully. In ships, as far as susceptibility to danger is concerned, there is no difference between the admiral, the cook, the engineer and the weapons officer. Finally air forces are again equipment-centric organisations, but in air forces the actual fighting is done by a very small number of people. In air forces the officers are sent off to risk death by the other ranks on the ground.

Proposals for jointery must recognise the barriers that these different cultures and relationships to equipment and combat capability can throw up.

JOINTERY AND MILITARY OPERATIONS

From a military perspective, most Western experience since the middle of the twentieth century has been that the branches of the armed forces live largely separate lives during peacetime and then are called on to act together when military operations begin. In a range of wars since 1939, the capacity of a state's armed forces to act as one was a significant factor in the state's overall success.

Increasingly the British Ministry of Defence has recognised that effective joint action benefits from careful prior thought and planning, and of course training. It has concluded that much single service training should be shaped with joint operations in mind. Even so, many areas of joint action remain problematic, including the integration of offensive air power with ground forces and ground-based air defences:

> The Government is determined to uphold the leadership, loyalties and traditions which are essential to the morale of the individual armed Services and their fighting capability. This country's experience of modern war, most recently in the Falklands Campaign, has progressively demonstrated, however, the need for the Services to be trained and equipped to fight together.[5]
>
> We will be conducting many more command, training and support activities on a joint Service basis because we expect almost all future operations to be joint.[6]

In British thought, joint action is facilitated by common doctrine and thinking about the nature of military capability. Military capability is defined as having seven components – Inform, Command, Prepare, Project, Operate, Protect and Sustain – and each service contributes to these capabilities in different ways.[7] The sensors and communications equipment aboard ships, for instance, are an important element in the UK's overall Inform and Command capabilities.

Two other central messages of central importance for each service are that UK forces need to be prepared for high-intensity combat, despite the incidence of peace support operations. Moreover all services need to be prepared for an operation of one character, for instance peacekeeping, to transform over time into another, such as peace enforcement.

Andrew Dorman and his colleagues have argued persuasively that the pressures for jointery have increased with the end of the Cold War.[8] The East–West confrontation involved large scale but brief and predictable military operations that could be planned in depth by armed services with separate identities. Since 1990, however, unforeseen force projection missions have dominated the activities of NATO forces. The flexibility needed for such missions can best be generated through deeper joint preparatory activities.

Moreover, since the mid-1980s, many military analysts in the US have concluded that a Revolution in Military Affairs (RMA) is imminent. The most prominent features of this anticipated RMA is that competitive military advantage will stem from the ability to collect and process information, and from the integration of sensors, communications equipment, command and control systems, platforms and weapons into a 'system-of-systems'. This system-of-systems will generate rapid and decisive action.

In terms of the basic relationships with technology mentioned earlier, even the infantryman is being viewed in some thinking not as person per se, but as a platform on which a range of capabilities can be based. Alternatively an infantry unit can be viewed as a platform. Such ideas have support in other European states, including the UK.

RMA and system-of-systems thinking is no respecter of single service divisions with thinking increasingly focused on capabilities rather than environments. As the term Revolution in Military Affairs implies, changes in military organisations, cultures and doctrine will be needed if full advantage is to be taken of the possibilities offered by new technologies. There seems little doubt that the RMA will bring pressures to further erode single service boundaries.[9]

After 1990 the UK found itself drawn into a series of military operations, especially in the Gulf and former Yugoslavia, that involved all three services. Until 1998, the UK response was to set up an ad hoc joint military headquarters, headed by a joint force commander, to run these operations. However the Government decided in its Defence Costs Study of 1994 to establish a Permanent Joint Headquarters (PJHQ) based at the naval headquarters at Northwood on the outskirts of London. This headquarters became the focus of the extensive communications and planning capabilities associated with the projection of British forces. In 1998, after the Strategic Defence Review, it was placed under the control of a three-star Chief of Joint Operations (CJO), who normally became the UK-based Commander of UK forces when they were deployed. However, it also became a central element in peacetime for the planning associated with the sorts of operations with which the UK felt it may become involved in future.

The UK has defined a pool of units and equipments with associated transport capabilities, known as the Joint Rapid Reaction Forces (JRRF). These are obviously available to CJO. These are units at high rates of readiness which can be deployed at short notice on a range of force projection missions. A medium scale operation for the JRRF would comprise a brigade group in terms of land forces, a maritime force of around 15 major warships, 60 fast jets and 50 other helicopters and aircraft.[10] The JRRF units remain on their home service unless they undertake joint training or operations. They may thus be seen as effectively cooperating rather than deeply integrated forces.

As a three-star officer, CJO is inferior in rank to the single service chiefs, to the single service front-line commanders (who are responsible for preparing force units at specified rates of readiness) and to the single service heads of personnel, all of whom hold four star posts. However, given CJO's central role in the preparation of joint capabilities for future operations, it is clear that he should be providing direction to the single services. This suggests that before long CJO will be defined as a four star post and his comparative weight (compared with single service chiefs) will be recognised.

Figure 1 summarises UK model for the joint command of operations under political direction. Political direction is provided by the Prime Minister and Defence Minister, who often work through a committee of political colleagues forming a war Cabinet.

They receive advice from the Ministry of Defence, where the Chief of the Defence Staff, who is the Prime Minister's chief military adviser, and the Permanent Under Secretary, who heads the civil service element, provides defence political input.

The Government oversees the military operations which are under the control of the Joint Commander in PJHQ. This is normally CJO himself. His immediate subordinate is the Commander who is deployed with the force.

FIGURE 1
UK MODEL FOR OPERATIONAL COMMAND

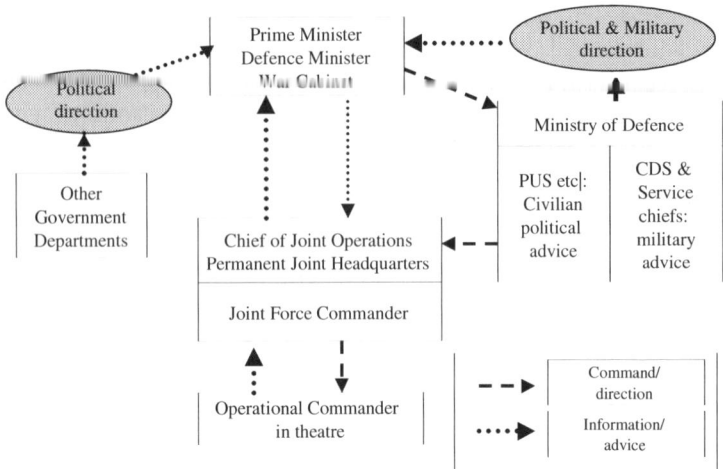

The dotted line linking PJHQ with Northwood is a symbol recognising that information flows may sometimes pass directly from the War Cabinet to PJHQ, bypassing the MoD. Especially in fast moving 24 by 7 operations, maintaining the MoD machine and PJHQ in synchronised operation presents real challenges. However, these stem mainly from the need for the political direction of military operations rather than from the joint nature of modern operations.

The joint command of operations cannot work well unless the groundwork has been laid in terms of the prior training of forces and commanders. The official and joint British Defence Doctrine shows why it is believed that there must be extensive preparation for joint operations and asserts the central ambition of joint activities - to derive greater military capability than could be obtained by having the forces act in isolation from each other.

> One vital key to the effective command of joint manoeuvrist operations is recognition of the relative strengths and weaknesses (both inherent and situational) of each component of the force and the playing to each of its strengths in support of the others. In doing this, the commander must concentrate on the effects he needs to

generate and employ the best possible means of achieving them. This will often require lateral thinking and the employment of units in ways not traditionally associated with their principal operating environment. By adopting an effects based approach to operations and utilising all the elements in an integrated fashion, the value of a joint force is more than merely the sum of its constituent parts. To achieve this requires an instinctive joint state of mind.[11]

JOINTERY AND MANAGEMENT EFFECTIVENESS

Military operations can be mounted only with forces that have been organised and prepared to undertake them and attention can now be turned to the management of the defence sector, including the preparation of forces, and to the pressures for jointery in this area.

Twentieth century armed forces have often shown a preference to be as self-sufficient as possible, making their provision for the supply of a wide range of goods and services from within their own organisational structures. Moreover individual service branches have rarely been keen to share support organisations, resulting in extensive duplication.

The equivalent tendency in the commercial world was the prominence in much of the twentieth century of the vertically integrated company, such as Ford in the 1920s. The management analyst Charles Handy links such desires for self-sufficiency with the territorial instinct found in many creatures, an instinct that drives them to seek to mark off an area as their own. Others will be kept out of this territory if possible.

> Groups have their own ways of fencing off their territories . . . They create their own language and their own culture . . . Groups need their territories. It is one way in which they define themselves as a group. But if those territories are too well protected, the group becomes an island, cut off from the rest of the organisation . . . Reorganisation inevitably means a redistribution of territory. This is perhaps the reason why it is disliked by so many . . .[12]

His observation summarises in many ways the problems of allowing single branches of the armed forces to run their own affairs. The Navy, the Army and the Air Force can become 'islands cut off from the rest of the organisation', that is from defence as a whole and the wider national interest that is the proper concern of the government.

For more than 50 years in Britain the pressures to diminish single service autonomy and self-sufficiency have been building as a result of public expectations that defence funds should be spent efficiently and effectively, and because of the increasing financial pressures on the defence budget.[13] There is no reason to think that these pressures will not apply in other states, indeed there are many signs that they do. The United Kingdom, however, may be thought of as a something of a special case because of the extensive defence capability, not least in the area of force projection, that it seeks to obtain from a modest defence budget. But as more countries feel pressures on their defence budget as equipment and other costs arise, and as more states contribute to the maintenance of international security through the use of their armed forces, lessons from the UK may be of wider interest.

FIGURE 2
TRADITIONAL MODEL

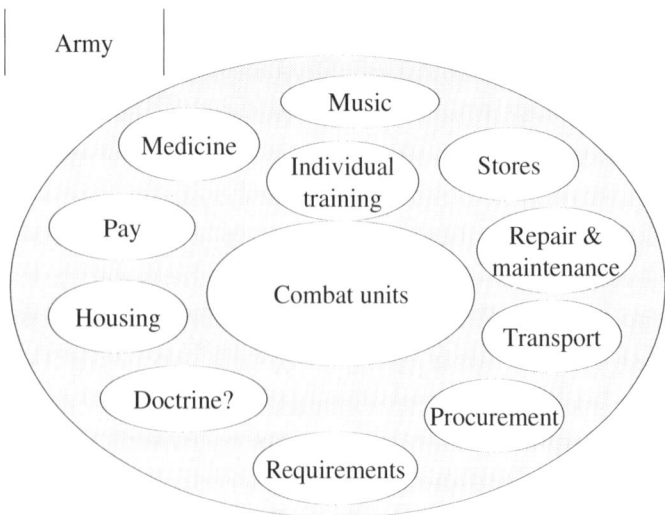

Figure 2 seeks to summarise the service self-sufficiency that was a feature of the UK in the years immediately after World War II. The British Army decided upon the requirements for its equipment and arranged their purchase. It looked after its own transport, repair and maintenance, and stores. Its troops' pay, accommodation and medical needs were met by army organisations. The Army trained its own musicians for its regimental bands. It took responsibility for the recruitment of its own troops. The other services likewise provided for themselves.

FIGURE 3
EFFICIENCY DRIVE: OUTSOURCING AND JOINTERY IMPACT IN THE UK

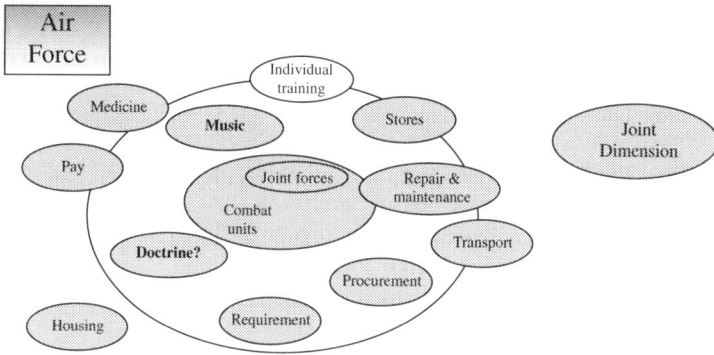

summarise the changes that have occurred so far as a result of the drive for efficiency and effectiveness. More and more activities have been contracted outside the defence ministry and the armed forces to the private sector. These include housing and many aspects of pay, transport, storage, repair and maintenance. Even individual training, including some combat-sensitive elements such as tank driving and basic pilot training, have already been entrusted to the private sector.

Many other aspects of defence life have been placed in 'purple' organisations, within the governmental defence sector, that serve all three services. There is a Joint Doctrine and Concepts Centre concerned with the high level doctrine within which single service doctrine must be accommodated. Individual staff colleges have been closed and a Joint Services

Command and Staff College established, particularly to deliver an Advanced Staff Course taken by a top 25 per cent or so of officers in all services.

Requirements are written and placed in priority order by a purple organisation, the 'Equipment Capability Customer'. All purchasing of operational equipment is done by the Defence Procurement Agency. Equipment support, except that needed on the front line, is done by the Defence Logistics Organisation. This is a huge organisation that employs 45,000 people as well as contracting for the services of many others. The musicians of the Army, Navy, Air Force and Royal Marines are now trained at a single school.

For reasons of economy, the UK has even created some joint forces, bringing together the assets of two or more services into a single organisation:

- Joint Force 2000 brings together the Sea Harrier FA2s of the Royal Navy and the RAF's GR7 Harriers into a single force of vertical and short take-off and landing (VSTOL) combat jets able to operate from airfields or aircraft carriers. Eventually these two aircraft types will be replaced by a single type, the Joint Strike Fighter.
- The Joint Helicopter Command draws on the assets of all three services to provide the optimum package to Joint Force commanders on operations. It directs the training, planning and resourcing of the Navy's Commando Helicopters, the RAF's support helicopters and operational Army Air Corps helicopters.
- Joint Ground Based Air Defence brings together the Rapier units of the RAF Regiment and the Army's Rapier and HVM units so that they can both defend either land or air forces assets.
- The Joint Nuclear, Chemical and Biological Defence Regiment brings together the assets of the Army and the Air Force to make available a force to be available for deployment at short notice.
- The Army and Air Force signal units have been co-located to form a Joint Service Signals Unit.

The creation of joint units has certainly produced major challenges, essentially because people from different organisational practices and cultures come together, and more work is needed before cohesion can be maximised.

The area of management thought that addresses such issues is 'change management' with an emphasis on the predictable nature of the reactions to proposed change, the importance of sustained leadership from the top, and the need for time to embed change. In many ways the creation of joint organisations presents similar problems to the merger of companies. In each case there is a need to unify differing formal mechanisms (expense allowances, pension schemes, information systems etc). But in addition, the different values and informal practices (i.e. culture) of the organisations have to be reconciled. This is a difficult challenge, as is widely recognised.

Moreover the economies from joint activities may not be as great as might be expected since joint programmes tend not to give up any of their elements. The courses at the Joint Helicopter School tend to be longer than their single service precursors because none of the differing features of the original courses has been given up.

The mix of joint and single service organisations within the UK presents challenges of coordination, as can be illustrated by consideration of what is needed to turn equipment into real military capability. Using formal UK thought, which reflects basic logic, equipment is not usable unless provision has been made for:

- Logistics support in the form of spares, fuel and so on;
- Training users and maintainers in the use of the equipment;
- The application of doctrine that specifies how the equipment is to be used, not least in conjunction with other equipment;
- The organisation of units that can operate and support the equipment; and
- The recruitment and basic training of individuals who will be able to operate and maintain the equipment.

In the UK, responsibility for delivering performance in different aspects of defence is allocated to a series of Top Level Budget (TLB) holders.

FIGURE 4
RESPONSIBILITY FOR THE PREPARATION OF MILITARY COMPATABILITY

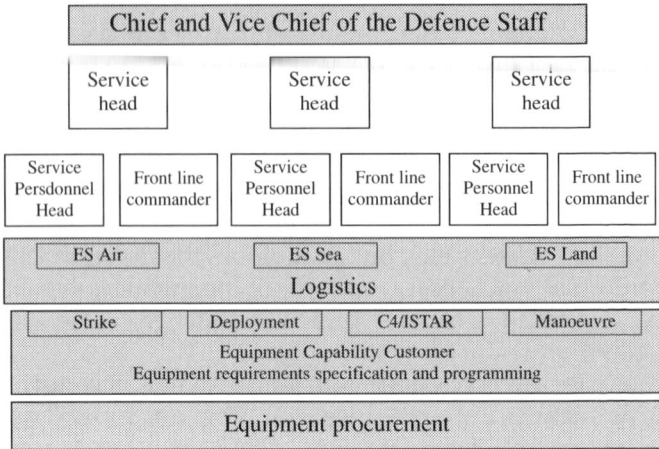

Beginning at the bottom, financial control over equipment spending is shared between the Equipment Capability Customer and the Chief of Defence Procurement who heads the Defence Procurement Agency. Responsibility for finding the money to support equipment once it comes into service is the duty of the head of the Defence Logistics Organisation.

Training the users and the field maintainers of the equipment is the responsibility of the 'front line commander' of the relevant service whose overall responsibility is the preparation of units of forces at specified levels of readiness so that they can be deployed if necessary by the government. Doctrine for the use of the equipment must be provided by the central staff of the relevant individual service, working with the guidance of the Joint Doctrine & Concepts Centre. The individual service heads are not TLBs and their activities are funded from the funding of the 'purple' (tri-service) Vice Chief of the Defence Staff.

Ensuring the availability of a suitable organisation within which to locate a new piece of equipment also is largely a service chief responsibility.

The recruitment of suitable personnel to operate and support new equipment is the responsibility of the personnel heads of the single services who, like the front line commanders, are TLB holders.

The smooth coordination of the plans and priorities of all these stakeholders cannot be taken for granted and indeed requires significant and continuous attention, as well as a shared commitment to making the system work.

JOINTERY AS A PROCESS

So far it has been shown how pressures for the effective conduct of operations and for the effective management of defence resources in peacetime make for increased jointery.

The next point to stress is that jointery needs to be seen as a process of continuous change, not as a condition that has or has not been achieved. When a change is made to capture the benefits of integration or cooperation in one area, it often leads to spillover pressures for cooperation and integration elsewhere.

This can be illustrated by reference to the case of the British government's choice, reflecting the ad hoc practices developed for the Falklands War (1982), that RAF Harriers as well as Royal Navy aircraft should be able to operate from naval vessels.

Superficially this just required that RAF pilots need to be trained to land their aircraft on moving as well as fixed landing sites. With VSTOL aircraft this was a minor addition to training needs. As this plan was implemented, however, it became clear that the RAF and Royal Navy had different personnel arrangements to support the aircraft, with the Navy training its maintainers with more skills, so that an RAF maintenance team for an aircraft was larger than its naval equivalent. This had implications for accommodation on board ship when RAF planes were deployed there.

It also became clear that the Royal Navy and RAF had different pay rates and allowances for people doing essentially the same job, raising pressures for harmonisation in this area.

Having owned the aircraft separately, the two services had developed different rules for the safe operation of the aircraft, which related to weather conditions, loadings, manoeuvres and so on. All these areas produce pressures for further change and indeed the MoD has now established a single

centre, the Defence Aviation Safety centre, to determine
safety regulations for all aircraft types. There is also a range of
issues arising relating to the safety regulations for the loading
and storage of munitions on ships.

Similar ripples of pressures for change have emanated
from the decision to create a joint school to train helicopter
pilots from all three services.

These more recent illustrations of change in one area
generating pressures for change in related areas should not
disguise the long-established trend in the UK towards a strong
central ministry of defence and erosion of single service
autonomy. No detailed history will be presented here but the
bare bones of information presented in Table 1 make clear that
a trend of this nature has been nearly continuous since 1945.
The chronology also shows the long history of separate
organisations for the different services and so suggests the
resistance that schemes for increased jointery can provoke.

Before leaving this summary of change in the UK, it should
be stressed that the movement towards the weakening of
single service autonomy and the rise of a stronger central
MoD has not meant that control over defence spending has
become more centralised. Since the 1980s the UK has
developed a management system in which users' budgets
devolved to a network of Top, High and Base Level Budget
Holders who are charged with the delivery of specified
outputs of capabilities and services. In a variant of the
thinking behind the military concept of Mission Command,
MoD budget holders are told what their chiefs want to be
delivered, and given some freedom in their choice of means
and methods of delivery.[14]

There is little doubt that jointery has some real
disadvantages that must be directly addressed. In particular,
those operating on land, at sea and in the air need their own
organisations at some levels and knitting single service and
joint organisations together cannot be straightforward. This
has been illustrated in this chapter by reference to equipment
acquisition but other areas could have been used.

Moreover, to move from a system dominated by rival
single services to one where joint activities have the lead
requires considerable effort and, in some cases, expense. As is
widely observed, organisational change is often unpopular

KEY EVENTS TOWARDS UNITED KINGDOM INTEGRATED FORCES

1414 First known Master of the Ordnance

1546 The Admiralty formed by Henry VIII to take administrative control over the Navy. This turned into the Board of the Admiralty

1660s Secretary at War's Office: first post holder killed at sea by the Dutch. This turned into the War Office responsible for the Army

1918 Creation of the Air Ministry

1940 Ministry of Aircraft Production which evolved into Ministry of Aviation Supply in 1970

1946 Ministry of Defence Act: MoD as a weak body of mainly civil servants seeking to coordinate the three services, each of which had its own Cabinet minister

1957 MoD expected to formulate defence policy and single service ministers lost Cabinet status

1958 Chief of Defence Staff (CDS) title established for the chair of the Chiefs of Staff Committee

1963 The Central Organisation for Defence established a unified MoD and separate service ministries abolished. The Defence Council became the highest level MoD group and two Permanent Under Secretaries were appointed to coordinate the business of the Ministry

1972 Ministry of Aviation Supply brought into MoD and unified procurement organisation was established (the Procurement Executive)

1981 Single service ministers abolished and replaced with junior ministers responsible for the Armed Forces (people) and equipment.

1982 CDS strengthened as Government's chief military adviser

1984 Abolition of single service policy and administrative staffs and responsibility for these issues placed in a central MoD Defence Central Staff. Service Chiefs made responsible mainly for preparation/capabilities of their services. Single service scientific advisers abolished and Chief Scientific Adviser strengthened

1991 The various research laboratories of the MoD brought together in a single organisation, the Defence Research Agency which later acquired responsibility for test and evaluation activities

1994 Defence Costs Study: establishment of Permanent Joint Headquarters
Joint services recruitment centres
Joint Advanced Staff Course to replace single service courses at a Joint Services Command and Staff Course
A joint Higher Command & Staff Course from a course run by the Army.
Joint Helicopter Flying School
Increased joint training in specialist activities (driving, foreign language, animal training, music etc.).
Stronger central direction under MoD Surgeon General overseeing tri-service medical provision

1998 *Strategic Defence Review:*
Standing Joint Rapid Reaction Force and three-star Chief of Joint Operations at PJHQ
Defence Logistics Organisation to address support on a joint basis
Equipment Capability Customer organisation established to set and prioritise requirements on a capability not a single service basis.
Creation of Joint Doctrine & Concepts Centre

1999 - Modernising Defence: including a centralised policy for information and stress on the collegiate responsibilities of the Defence Management Board (four-star civil service and military body including the individual service chiefs)

and the Ministry of Defence has always needed to take care not to change so rapidly as to erode morale in the uniformed services. The management of change takes time and effort: those contemplating addressing change in their own states may wish to look in more detail about how long the UK has taken to change its system to where it is today.

INTERNATIONAL IMPLICATIONS

This chapter has been consciously UK-centric and thus needs some attention to be devoted to the activities of other states. Given the number of states in the world with at least three branches of the armed forces, argument is restricted to three broad points.

First, when countries seek to change too rapidly without building support for a new system, change fails. Canada decided to abolish its separate army, navy and air force in changes introduced in 1964 and 1968. It then realised that morale was suffering with the loss of association with a particular service. Rules were relaxed and, in 1985, it reverted to the wearing of separate distinctive uniforms for air, land and sea troops. Canada has, however, not abandoned its top level unified force.

There are parallels again with the area of European integration. The European Defence Community and European Political Community plans of the early 1950s collapsed in 1954. These changes would have created a single European foreign and defence policy at one stroke and they failed. However, they were succeeded by the more gradual change of the European Economic Community, the European Union, the Treaties of Maastricht and Amsterdam and so on.

Second, the United States is constitutionally oriented towards the separation of powers in Government. Just as the President and the Congress compete for power, so do the Army, Navy, the Marines and Air Force compete for resources. However, the US Government is aware of the needed for the coherent direction and operation of all its servicemen in military operations and its system of autonomous service branches is being moderated through the powers and the roles of the Joint Chiefs of Staff and their influential Chairman and the role of joint commands. The Defense Re-organisation Act of 1986, often referred to as the Goldwater-Nichols Act, places the responsibility for coordinating the activities of US single services on the Chairman of the Joint Chiefs. Increasingly officers know that, for promotion to senior positions, they need to have done a tour in a joint organisation.

The US is driven more by the operational military benefits of jointery than by the management-based pressures that affect countries less well endowed with defence resources. The operational pressures are growing, especially as the US is placing such stress on information-based warfare. There is thus increasing political interest in promoting jointery in America. However, the traditionally autonomous services struggle to integrate or even coordinate their activities.[15]

Third, jointery is advancing steadily in states that are pressed for defence resources. Most states already have 'purple' organisations that handle all their operational procurement. In France the Délégation Générale d'Armement (DGA) has long been very powerful, in defining both what should be acquired and how it should be bought. In Australia the recently-formed Defence Materiel Organisation both procures and supports equipment for all the armed forces. Germany is introducing a single organisation to provide for the support of all its services and is also strengthening its central operational staffs.

In short, the global trend is towards greater joint direction and management of armed forces, driven by the operational pressures and the needs of effective management.

To look forward also leads to the conclusion that jointery among branches of the armed forces is far from the end of the story.

The pressures for the effective management of defence resources will result in more defence-related tasks being placed in the hands of civil servants and even private contractors: there is little need for most financial, personnel, contracting or catering positions to be filled by professional soldiers who are expert in the organised application of armed force. Arguably civil officials should also have an important place in the generation of a defence policy that is linked to foreign policy and other areas of political concern.

This leads to the conclusion that the effective direction and management of defence policy cannot thrive where there is a 'them and us' mentality between civil servants and military staff. Twenty-first century jointery thus needs to address the linking of the three armed forces and the civil service – what this author likes to call the purple pinstripe dimension (a reference to the traditional pinstripe suit of the civil servant and the purple of that which is traditionally used to signal a tri-service/joint activity).

Finally the contemporary armed forces of most states, when engaged in operations, are often involved in affairs much more complex than 'simple ' battles and the securing of territory against hostile, conventionally organised forces. They are one (important) element in the political challenges of defeating insurgent movements and in conducting peace support operations. These activities require the integration of armed forces' actions with those of other government departments. The British Conflict Prevention Fund has been established to coordinate and integrate the expenditure in this area of the UK MOD, the Department for International Development and the Foreign and Commonwealth Office.

Also, on peace-support and combat operations, armed forces have to work alongside the forces of other states. Especially in peace support operations national military forces come under pressure to work effectively with non-governmental organisations (NGOs) such as Medecins Sans Frontières, Oxfam and so on. UK forces in Sierra Leone, for instance, are contributing to recovery in a country where dozens of NGOs are also active.[16]

European political integration is often analysed in terms of deeper and/or wider. Deeper involves more intense integration among existing EU membership, for instance beyond a common currency to common company law and common taxation systems. Wider involves extending the European Union to cover more states. Jointery for armed forces has a 'deeper' dimension where the separate branches of the armed forces move towards, for instance, common pay and allowances. But the wider dimension, involving military links with civil servants, other government departments, other governments, and even non-governmental organisations, should not be overlooked.

REFERENCES

1. For a discussion of the definition of jointery, see A. Dorman, M.L. Smith & A. Uttley, 'Jointery and Combined Operaitons in an Expeditionary Era: Defining the Issues', *Defense Analysis*, Vol.14, No.2 (1988) pp.1–5.
2. Martin Edmonds, stressing jointery as neologism, stresses that jointery can refer to activities undertaken by military personnel and civil servants, see 'Defense Management and the Impact of Jointery', *Defense Analysis*, Vol. 14, No. 2 (1988) p.10.
3. 'Spillover' as a concept has moved into and out of intellectual fashion but, with the implementation of the Single Market programme and the introduction of the Euro, appears to be assured of a respectable and lasting place in thought about European integration. Two recent works addressing the issue are N. Nugent, *The Government and Politics of the European Union* (New York, Palgrave, 4th edition, 1999) pp. 507–8, and M. Williams, *International Relations Theory and European Integration* (London: Routledge 2000). For an older work, see Michael Hodges, 'Integration Theory' in T. Taylor (ed.) *Approaches and Theory in International Relations* (London: Longman 1976) pp.237–56.
4. British Defence Doctrine, *Joint Warfare Publication 0-01*, Shrivenham, Joint Doctrine and Concepts Centre, October 2001.
5. UK Ministry of Defence, *The Central Organisation of Defence* (London: HMSO 1984) Para. 5.
6. Secretary of State for Defence, Malcolm Rifkind, 1994.
7. See British Defence Doctrine (note 4) pp.4.2–4.3
8. Dorman *et.al.* (note 1) pp.1–8.
9. For a discussion of the relationship between jointery and a Revolution in Military Affairs, see B.R. Sullivan, 'The Future Nature of Conflict: A Critique of "The American Revolution in Military Affairs" in the Era of Jointery', *Defense Analysis*, Vol.14, No.2 (1988) pp.91–100. He observes: 'to be more effective in winning war, land power can be augmented by proper co-ordination with sea and air power. In other words, jointery is essential for sea and air power and highly useful for land power. This will remain true....' (p. 97).
10. Report by the Comptroller and Auditor General, *Exercise Saif Seria*, London, National Audit Office, 23 July 2002, p. 7.
11. British Defence Doctrine (note 4) p. 3.8.
12. Charles Handy, *Inside Organisations* (London: BBC Publications 1992) pp. 50–1.
13. For a wide-ranging summary of defence management reforms in the UK and the place of jointery within the, see Edmonds(note 2) pp. 9–28.
14. See www.mod.gov.uk (How the MoD is organised) and Ministry of Defence, *The Organisation and Management of Defence in the UK* (London, MoD, undated).
15. See an authoritative digest of the report of Retired Vice Admiral A. Cebrowski to Secretary of Defense Donald Rumsfeld ('Disjointed First Steps') in *Defense News*, 19–25 August 2002.
16. The coordination let alone integration of the activities of NGOs is in itself a major problem. Joining up the activities of governments and NGOs, even when they share common aims, adds a further level of difficulty. See for instance, A.M. Fitz-Gerald, 'Setting the Scene: the New Conflict Environment and Callenges for Future Interventions' *Conflict, Survival and Medicine*, Vol.18, No.1 (Jan.–March 2002) pp.5–23 and A.M. Fitz-Gerald, P. M. Molinare and D.J. Neal, 'Humanitarian Aid and Organisational Management', *Journal of Conflict, Security and Development*, Vol.1, No.3 (December 2001) pp.135–145.

5

Professional Armed Forces: Concepts and Practices

PATRICK MILEHAM
University of Paisley, Scotland

PROFESSIONS AND PROFESSIONALISM

'A profession', noted Samuel Huntington in 1957, 'is a peculiar type of functional group with highly specialized characteristics.' He continues, 'Professionalism distinguishes the military officer of today from the warriors of previous ages.'[1]

Armies, Navies and Air Forces are organisations[2] whose purpose has long since shifted from the relatively straightforward, practical function of fighting, to an enormously complex institutional matrix of processes and procedures, carried out by numerous inherent and out-sourced resources, human and material, to achieve certain ends. While Huntington was writing specifically of commissioned officers, another writer, Jacques van Doorn, wrote in 1975 of 'The officer corps: a fusion of profession and organisation.'[3] The implication, in the mid years of the Cold War, was that the officer corps personified the contemporary standards of a professionalised and systematised leadership of national armed forces. The corporate body holding commissioned rank and full legal authority, organised the monopoly of violent means, reserved by the state for its defence.

Significantly, this accords with international law, which requires soldiers 'to be commanded by a person responsible for his subordinates',[4] a statement which establishes both authority and subordination. In another context, that of military public relations and career advertisement, for several years the British Army was promoted as 'The Professionals',[5] the implicit message being that the Army was a serious and responsible institution and that the term embraced all its members, not just commissioned officers. 'Professional' was used to counter the term 'amateur' and lingering shades of the illogicalities and inefficiencies of the National Service experience.

While other chapters in this book analyse the human agencies that constitute so-called 'Post-modern' armed forces in respect of the supposedly necessary policies and practices, this chapter has a different purpose. The aim is to explore, rather than to define, the concepts and practices of professional armed forces. This includes much more than a study of the officer corps. While in any definition there are obvious quantifiable components, both material and human, the *quality* of the characteristics of professions and professionalism is my subject. The supposition is that there is no contemporary (2003) qualitative documents and list of criteria which express professional concepts and practices to assist in the reform of armed forces in modernising nations – notably of the former Warsaw Pact – let alone those advanced nations displaying characteristics of post-modernism in their armed forces, (whatever that expression means).

Another contention is that the very notion of post-modernism[6] may actually undermine theoretical and practical understanding of human quality and values, by introducing misleading category errors, false assumptions and dysfunctional reforms. The chapter thus is intended to establish enduring or permanent truths and realities about 'the profession of arms'[7] which will always be *contemporary*, whatever are the pressures to accept a *future-now* mentality. The quality of general professional occupations and professionals as people, is the starting point of this chapter.

Originally a profession was an exclusive group of persons specialising in an occupation, widely claiming, and more or less firmly establishing, certain rights to practise their services or provide goods among the general public. The 'contrast' with other occupations and professions, and the 'authority'[8] assumed and accepted, gave and continues to give those specialists particular recognition, legitimacy and in many contexts legal status. 'Professional autonomy' is asserted, the result of knowledge bringing power, manifest in 'knowledge-based competence'[9] and delivery. Power is thus vested in individuals, legitimated by institutions and the wider professional constituency,[10] the implication includes acceptance of individual and corporate responsibility.

Close reading of the literature of professions, professionals and professionalisation adds other definitions. With direct remuneration for services provided, as well as growing social accountability, professional people have always recognised that their services are of value and need to be valued, respected, even honoured. The early history of the professional institutions, such as guilds in the medieval context, reveal imposed disciplines and self-disciplines for the protection of value and promotion of value-recognition.

With the bureaucratisation[11] of professions, by definition, they now include comprehensive codification of practice, establishment of standards, often standardisation of practice, as well as personal codes of professional conduct. Furthermore the growing democratic and 'social control of expertise',[12] as opposed to the professionals' sole control of expertise, is a particular feature of the second half of the twentieth century in many countries.

Professional bodies, with legal or quasi-legal status, are empowered and expected to certify and enrol those joining the profession, often after considerable periods of professional education and training, with rigorous examination and assessment of competence. They also have the responsibility for punishing and disbarring professionals who fail to maintain standards expected by the professional body and wider society. Professional institutions initiate, support and encourage the furtherance of specialised and general knowledge and understanding through comprehensive academic study and developmental research, in conjunction with universities and research institutions. Most professions have a large and constantly growing corpus of published doctrine and research.

In most advanced liberal democracies, most occupational groups contain large numbers of fully-qualified or quasi-professionals – working in commerce or private practice, the public sector and the third sector (a mixture). This is the extent to which professions are now bureaucratised and open to public scrutiny – not wholly self-authenticating, self-regulating and unassailable in their power, as once they were. Thus the final responsibility of modern-day professionals through their institutions, has become increasingly a matter of an intensification of relations with the public.

ARMED FORCES: NATIONAL INSTITUTIONS

In the extended introduction above, the nature of national public sector institutions has already been established in part. I believe that the first group of qualitative defining and enduring characteristics of professional armed forces are:

1. Objective control by the civil authority,
2. Relationship with all national security agencies,
3. Relationship with the civil population, and
4. Quality of force design and resources, to match expected capabilities for defence needs.

I shall take each in turn.

In modern liberal democracies, all defence and security forces and agencies are under the command, legal control and policy direction of constitutionally recognised and democratically-mandated public appointment holders, and officials. The definition of the

> holding of an office…[embodies] expectations of….certain standards…of the agent or office holder. The office is a trust in the legal sense of trusteeship,'[13]

in this context public trusteeship.

In democracies, therefore, executive power is based ultimately on elected trusteeship. Absolute, permanent power is categorically denied to individuals by the electorate. There is, however, constitutional separation of powers between elected representatives and independent, permanent[14] professional civil and military servants of state. This does not mean that unelected professionals hold constitutional powers over elected representatives, in any other respect than trusteeship in their professional capacity, defined or implicit. In modern democracies, by definition, military and civil officials are subservient to elected representatives as agents of state control. Legally and by custom they are excluded from the 'political process'. Military intervention in politics, or failure to comply with constitutionally legal policies, direction, orders and instructions given by ministers (or civil servants acting under the authority of ministers), lay military officers individually open to charges of acting illegally under military or criminal law, or the lesser charge of unprofessional conduct. Acts of national parliaments, however, need to be explicit, reflecting the separation and subordination of military activities to civil control.

Normal customary procedures and processes should clearly demonstrate the professionalism of as well as define the corporate professions of politicians, civil servants and military officers. For routine business, however, it is neither sensible nor desirable that absolute control over all, or even much ,of the detail of military resources and activities is maintained by elected politicians, government ministers and permanent civil servants[15] – or indeed other civil institutions, agencies or organisations. The technical details of forming, maintaining, equipping and manning armed forces, the intricate routine direction and management of specialist military ways and means and conduct of events, should be entrusted to objectively selected, enrolled, trained and formally appointed commissioned and non-commissioned officers. They need to be clearly differentiated from civilians by their formal appointment and authority, as well as the visible custom of wearing uniform, with individuals' relative status confirmed, legally and symbolically by distinctions of rank.

The quality of the civil-military relationships is thus based on degrees of trust among trustees, as defined above. Each category of official, political, ministerial, civil service and military, has duties within this relationship. It is essential that ministers should not formulate rash or unreasonable policies, or give reckless commands, orders or instructions, either in the routine administration and management of armed forces, or in operational use. Military officers in truly professional armed forces do not have the right to refuse the call to comply with orders, but as a professional duty, they have the right to use every means (except political) of persuasion to prevent the launch of reckless and ill-considered military actions and operations. Joint routine professional work, including risk-calculation, with respect to the expected end-state of military action, linked with diplomatic, economic and other considerations and actions, should reduce or eliminate professional (or even political) tension between categories of professionals.

Explicit understandings about the military function is also necessary. For instance, British Defence doctrine draws distinction between various qualitative components of Fighting Power and hence overall 'Military Effectiveness' of the armed forces. They are the quality not only of decision-making and implementation, but the pre-determined,

• Conceptual Base,
• Moral Base,
• Personnel Base,
• Materiel Base, and
• Supporting Infrastructure.

The term 'professional' is manifestly a combination of the conceptual, personal and moral components.[16]

Good civil-military relations is thus a qualitative expression of good faith among all professionals, as well as the public who support the professional categories cited above. If civil-military relations are of low or declining quality, willingness of officers and servicemen and women to join and remain in service is affected in the medium to long term, if not in the short term. Whether conscripted armed forces by definition can be termed professional, will be argued later.

While the above paragraphs have been couched in contemporary terms, rather than post-modernistic, none of the comments so far need revision to fit post-modern notions, with the exception of internationalisation or globalisation arrangements for armed forces agreed by governments. In respect of armed forces working in alliances, the same high quality of relationships between elected authorities, civil executives and other national armed forces needs to be promoted.

In the future, should nationally mixed electorates become normal and mixed military units[17] be formed – countering the present status of national 'subsidiarity' in the European Union for instance – such matters a civil-military control and relationships may become problematical.

In the above remarks, certain other assumptions have been made which need further explanation. Democratic, constitutional arrangements and elected governments exist to promote the security of the nation and the freedom of individuals to act within legally defined and reasonable constraints. Security, however, is a much wider term than the provision of defence against external aggression. Armed forces exist to provide for this contingency, but they may have other roles, both internal such as military aid to the civil authority (e.g. firefighting and fuel delivery in times of strike) and in external contexts (e.g. defence diplomacy and training support to other nations).

Similarly, other ministers have responsibilities in regard to state security, and by association, physical defence such as the Home Office, in respect of civil defence. It has to be added that the quality of all national constitutional arrangements impinge on the quality of the armed forces, both directly and indirectly.

The second characteristic of professional armed forces is rather more straightforward than the first, albeit depending on the constitutional position and political direction of other security and defence institutions and agencies. Typical national institutions are:

- Civil police (unarmed police),
- Gendarmerie (armed police),
- Militia, National Guard (volunteer),
- Citizen army (conscript),
- Volunteer, part-time armed forces, formed as military units,
- Border Guards,
- Customs officials (including immigration control and revenue collection),
- Other emergency agencies.

These are distinguishable because in most, democratic nations, the term 'Regular' is normally applied to full-time armed forces as the ultimate institutions for the monopoly of violence under government control. It can be argued that if one or more of these institutions fail in times of national emergency, the armed forces can be tasked in their place.

Some of the pre-determined arrangements in respect of these institutions are necessary for the armed forces to continue to act professionally, when they act alongside or in their assigned place. Every institution and agency should have clearly defined roles and routine tasks, with as tight parameters as are reasonable and workable. If there are routine and normal overlapping roles and tasks, they must also be defined.

It is then the quality of both normal and during-emergency relationships between the armed forces and other security and defence agencies, as well as with their directing and controlling authorities, that indicates the quality of overall professionalism. Again, the customary and continuing 'moral base' of all activities, induces military and civil institutions to act together in good faith. Most readers can bring to mind numerous examples of good faith and good practice, as well as instances of failure,[18] in modern democracies.

In regard to the first two indicators of professional quality, it seems to be a phenomenon within modern-day, 'post-modern' societies, prompted by elected politicians, that public service professionals have to work very much harder to maintain high levels of public trust than hitherto.[19] Indeed it seems that professionalism is everywhere declining, even in armed forces. That is unlikely. The professions have to guard against 'deconstructionism' and 'relativism' in concept and practice, and it requires researching evermore deeply into professional practice and the way they conceive, both of themselves, and the public they serve.

It must be added that liberal democracies have peculiar ways sometimes of undermining the very quality of their democracy, by over regulation, legislation and litigation.

The third indicator of military professionalism, is the relationship of armed forces with the civil population as a whole. There are of course, a very large number of variables – dependent and independent – historical, political, economic and demographic, both quantitatively measurable and qualitatively observable.

The relationship in individual nations is often measured and assessed, crudely or scientifically, with the volunteer enlistment figures or volume (percentage success) of the intake in countries where conscription still prevails. The European Values Group Surveys (Gallup), conducted in 1979 and 1989,[20] and the National Pride Survey (Chicago) of 1998,[21] have measured populations' views of their armed forces' popularity over the years. Routine or occasional internal surveys, such as the Tickner research,[22] provide evidence of expectation of persons on joining armed forces, directly from civilian life.

The relationship of the armed forces and the population from which they draw their members (including potential officers), depends on six chief variables, most of which are self-explanatory:

- How 'close' or 'isolated' are the armed forces from their society in spirit, attitude and 'visibility'.
- Voluntary or conscripted service.
- Martial or militaristic style of internal relationships within the military.
- Demography of age, qualification, promotion system and opportunity.
- The national/international labour market.
- Armed Forces' reputation.

All are inter-linked and inter-dependent. One needs special comment.

The demography of voluntary regular armed forces is a specially sensitive matter and one not generally understood. As navies, armies and air forces are action-based enterprises with 'offensive' and not just 'defensive' roles, they chiefly attract young persons at the beginning of their working life. As people mature they require more stability and seek opportunities to progress to more financially rewarding and intellectually or enterprising phases in their whole life career. To maintain vigorous armed forces, only a small number, probably less than one quarter, are likely to be retained by the armed forces as senior non-commissioned and officer cadres – the 'career' armed forces. The large majority leave, more or less bored, frustrated by lack of promotion or lack of it, or unpromotable. Many of them joined in the first instance to gain valuable transferable professional[23] or semi-professional skills.

The fully professional 'military' officers are thus likely to be only those who reach and progress beyond a certain status.[24] Thus, de facto, and revealed by demographic analysis, a nation can have a 'career army' and a 'conscripted' or 'voluntary national service army', to label the dichotomy, all within the single army. That is not to say that the two latter types are not professional in character and performance – if the persons are of generally high motivation, perform to high standards and bring success to the enterprises conducted by their particular armed force. This leads neatly to the fourth characteristic of professional armed forces.

'Force design' of armed forces is a structural and physical conception, subject to the close direction of policy-makers – both civil and military – matching role and tasks with capability. One starts with the constraints or opportunities of 'polemity...the ratio of the energy employed (directly or indirectly) in warfare or preparations for it, to the total amount of energy available to society.'[25]

Crudely this means the amount of gross domestic product set aside for defence and more generally, security. If an insufficient budget is available to man, equip and support armed forces to meet foreseeable or unforeseeable emergencies, they are likely to fail. Capability will not match tasks. Armed forces that obviously fail will be diminished professionally in their own understanding, and externally among populations, both home and abroad.

There is little space in this chapter to discuss force design in detail. Included, however, should be policy direction, management, funding and other resources to provide for a sufficiency of 'combat' units, supported directly and indirectly by 'combat support' and 'combat service support' units, and personnel (to use NATO phraseology). Systems, both material and managerial, together with processes and procedures, are included in force design based on sustaining operations over distance and time.[26]

The energy employed together with the design, development, procurement, maintenance and support of all material, together with relationships with home and international defence industries, are also closely connected with the quality of defence output. Any may cause operational failure in part, or in whole. Professional armed forces need professional infrastructures of sufficient comprehensiveness, sophistication, and flexibility to act quickly in emergencies and for as long as necessary.

In summary, armed forces cannot be properly professional, unless they have professional direction, resources and support from civil institutions and agencies, as well as the support and respect of the civil population. It will be apparent from other chapters in this book, that many of the variables are undergoing radical change, which can be labelled 'postmodern'. The military's relationship with society is certainly affected: this requires positive action to strengthen the civil-military relationship. Similarly the adaptability of force design to maintain highly capable armed forces is essential, if new, even 'post-modern' physical, conceptual or moral threats are to be successfully countered.

PROFESSIONAL POLICIES AND PRACTICES

The second group of qualitative professional indicators moves beyond military power and force capability, measured routinely in documents such as the well known SIPRI *Year Book*, IISS *Military Balance* and the RUSI *Index of Martial Potency*.[27] Significantly the latter acknowledges that merely comparing manpower and weapon holding '... does not take account of a nation's efficiency in using its defence resources, nor such crucial issues as morale, nor the types of capability that are maintained'.[28]

The next four indicators are thus internal and reflect policies and people. They are:

5. Doctrines, policies and management infrastructures,
6. Voluntary service,
7. Disciplines and self-discipline,
8. Education and training.

Doctrine singular and doctrines plural, are both mental conceptions and actual publications, which identify the specialist nature and particular practices of the profession. The quality of the doctrines expressed and in use, can also be assessed in a much wider general context: how they fit in with the other profession's doctrines and practices and how generally useful they are in terms of theory derived from practice and vice-versa.

Publications of policy and practice need to be written at various levels. In terms of combat, theory and practice, these levels are 'grand strategic', 'strategic', 'operational' and 'tactical' in NATO terminology. A large number of documents need to exist to reflect the complexity and sophistication of a nation's armed forces, and their activities and duties. Operational and tactical factors inevitably change: strategy is strategy that is comprehensive and comprehending, so strategic level doctrine changes less fundamentally, because good strategic thinking includes adaptability and flexibility.

Military doctrine proper, like any professional doctrine, has the purpose of providing the 'first principles', both empirical and inductive, from which the profession and its supporting institutions and agencies can, literally, conduct research for new details, or sometimes, new first principles. There is much evidence that many nations' armed forces have indeed developed into 'learned professions', in the sense that the law and the church have been so acknowledged for centuries.

Thus, quite distinct from the managerial infrastructure which is part of force design, professional armed forces need to maintain strong links with academic institutions, and to create internal institutions which can study and develop the armed forces, as well as their links with other professions, occupations and organisations. Some nations have military universities and specialist colleges. All have staff colleges, where specialists join together to learn the art and science of generalist thinking and the application force. Officer academies and training institutions for non-commissioned ranks and recruits, likewise exist in all nations with any claim to modern professional, armed forces.

The sixth characteristic of professional armed forces remains an open question. By definition, need fully professional armed forces be wholly 'voluntary'? On the face of it, it appears that there is intuitive movement towards this being a defining requirement in liberal democracies. Most nations in Europe, including former members of the Warsaw Pact, have discarded even limited conscription already, or are eliminating it over time with an expressed end-date. One cannot, however, say that the Israeli or Swiss citizen armies are less than professional.

In many respects this indicator of quality is closely related to the armed forces' relationships with the civil population, controlling authorities and other agencies of national security. To answer the question, one has to search among the psycho-philosophical complexities of individual and group motivation and morale, but common sense indicates that one volunteer, motivated by choice, may be worth several pressed men, as the familiar expression has it. One has to ask, however, how willing is the volunteer? Certainly, in the British military doctrine of 2000, there is the requirement for all members of the armed forces to accept the '...legal right and duty to fight and if necessary, kill, according to their orders and an unlimited liability to give their lives in doing so. This is the unique nature of soldiering.'[29] The question must be, is this the most significant of all indicators of military professionalism? It is at the heart of any professional 'military contract'.

The answer may be that the majority of those who join the armed forces voluntarily in the first place, are more likely to accept this unique liability than those who are coerced into joining. In different nations, different armed forces and indeed different parts of an armed force, a variety of factors exist from the beginning and before operations start. In the event of 'active service' or once combat begins, who knows how he/she or others will behave and act? Most of the evidence is *ex post facto*.

On balance, with the evidence at hand, I believe that fully professional armed forces should by definition be voluntary. This probably reinforces post-modern trends in society, although the same conclusion may be reached by different paths.

The seventh defining indicator follows closely. All professions are disciplined groups of persons, in many senses of the word. Military discipline, traditionally understood, means, according to Max Weber[30]:

> the consequently rationalized, schematically trained and accurate execution of received orders – without giving expression of personal criticism – and the constant inner submission to that objective.

Traditional and un-modernised armed forces rely on coercive conditioning by authoritarian, militaristic means. Modernised armed forces could be said to promote rational, enlightened, means to inculcate discipline. Volunteer entrants arguably respond more readily under enlightened means of induction, than militaristic.

Military laws, written or unwritten codes and practices of conduct pertain to national traditions, more or less modernised to suit changing conditions within a particular armed forces and society. Disciplinary procedures for wrongdoers are needed to show exemplary justice, punishment and to reform the individual. For serious offences, courts martial try individuals in similar, or not so similar, fashion as national civilian courts try cases of criminal or civil law. Other professions also have their own disciplinary procedures. The higher purpose of such procedures is to uphold high standards of professionalism, promote successful military endeavour and maintain public confidence.

A moot question arises if the 'self-discipline' is a higher, more voluntaristic, psycho-philosophical motivation than 'imposed discipline'. In voluntary armed forces, self-discipline certainly is highly prized as a natural consequence of voluntarism. The question is how necessary and what extent is imposed discipline able to develop inner, self-discipline? The answer revolves around Weber's expression above – 'submission'. In many respects, there can be no conclusive answer, because one has to consider the workings of the brain and the connection with the mind of each and every unique individual who joins, whether volunteer or conscripted.

The justification for induced discipline by authoritarian means (lightly applied for the intelligent, more strongly applied for the slow-witted) in the armed forces of liberal democracies, is to do with setting high standards. Be a recruit ever so well motivated and keen to show self-discipline, at the outset he or she may not be aware of the required institutional professional standards. The intelligent instructor will therefore quickly recognise high motivation and use appropriate methods to teach the potentially highly self-disciplined, intelligent recruit.

There should be no need to resort to highly militaristic methods which, for volunteer armed forces, are increasingly dysfunctional as time progresses. A post-modern approach to discipline, however, may actually result in lower standards of self-discipline being achieved, to the detriment of military effectiveness. Conversely a post-modern approach, if it enlightens the connection between ways, means and ends in individuals, may actually result in higher standards.

The eighth indicator of professionalism is directly connected with ways and means.

Training in the British Armed Forces has a history going back well over 200 years. Training is the improvement of skills, both practical and procedural when working with others.

Education in the armed forces is of more recent introduction. Ideally it is conducted for the improvement of the mind's capacity for understanding what, how and why things happen and why people act or should act in particular and general military contexts.

It is therefore sufficient to state that professional armed forces are defined by the quality of their training and education. For the career, as opposed to the short-service personnel in armed forces, increasingly weight has been placed on education, taking personal understanding and skill much above the next level of promotion. Only one generation ago in the British Army, it was entirely accepted that 'training [was for] people only in the skills they need, as near as possible to the time they are going to need and use them.'[31]

Education for the higher direction of forces, requiring independence of thought, ability to analyse critically and argue both orally and in writing, is now widely accepted in national armed forces as a necessity, if everyone is to do their job expertly, including enabling civilians to understand military activities. The proportion of university educated officers is a significant indicator of quality, although additionally they need to be able to think and act quickly.

Space precludes further discussion about the extent of the facilities and range of training and education opportunities necessary for truly professional armed forces. Extending the military education fields, however, has required an increasing amount of defence and security related research and development, to keep military officers (commissioned and non-commissioned) abreast of other professionals, and relate them with the civil population and other parts of the employment market.

If post-modern trends are detrimental to the military or any other profession, it is likely to be the result of the fragmentation, through over-specialisation and dangerous isolation of professionals, living in a 'bound-rationality' limbo of their own. Alternatively, radical new training and educational ideas and facilities may enhance armed forces' capabilities and professionalism.

PROFESSIONAL PERFORMANCE

The third and final group of indicators of professionalism is to do with defence output in general and corporate competence or armed forces in particular. Three flow easily is sequence, the fourth is more difficult to define. They are:

9. Technical practices and expertise, applied to success in military exercise and operations,
10. Officer-NCO-servicemen's (and women's) relationships,

11. Morale, Leadership , ethos and reputation,
12. Evaluation of efficiency, validity of measuring effective-
    ness.

Technical or specialist policies and practices are, of course
derived from doctrine, research, education and training as
passed on to new generations of servicemen and women, who
in turn develop new policies and practices. Added to which the
corporate wisdom of the armed services needs to be fostered
and the ability to perpetuate and regenerate that skills useful in
preparation and conduct of further military operations.

The armed forces are unusual in one sense. They spend
much time in training, preparation and exercising their skills,
both physical and mental, in only partial-expectations of
having to apply them in real military situations. Ideally most
wise people would live in the hope of not having to go to war
and not having to be involved in much danger or risk in
operations other than war (OOTW). Volunteer servicemen
and women seek degrees of 'adventure', if not danger, as a
strong motive for joining.[32]

Members of fully professional armed forces, however,
accept the policies, practices, preparations and peacetime
exercising of their skills, in the expectation that they are
thereby reducing the danger levels of real operations, where
combat and conflict is likely. If in some of the situations that
they find themselves, they appear to be prompted by post-
modern tendencies of their antagonists – for example,
terrorists – then training and exercising need to prepare
members of armed forces as appropriately as possible. Success
in new types as well as on-going operations is thus a measure
of the quality of professionalism.

Mention has been made above about military authority
and responsibility, as well as the holding of office. Force
design and management infrastructure also incorporates the
command structures and hierarchy. In military operations
and service generally, there seems to be a need for the
continuing dichotomy of those who give an order and those
who obey it. The commander may be a general, the obeyer
may be a colonel: in another context a corporal may be the
commander and the lance-corporal the man who obeys. Thus
all armed forces continue to employ a wide range of rank
distinctions among their members.

They also maintain formal distinctions between commissioned officers, non-commissioned officers and holders of no office at all – private soldiers or enlisted men. In their force design, individual nations need to categorize the status and the internal demographic shape of their armed forces personnel structure.

What can be stated about all members of the military hierarchy, is that there are formal, professional relationships at many levels, which define the hierarchy and determine who can make what decisions operationally – and in other contexts – and who has to obey them. Some of these formal relationships are well defined, highly developed and entirely relevant: others may be accidental or archaic. Clarity about who can impose disciplines and punish according to armed forces' law, is the crucial division between commissioned, warrant (quasi-commissioned), non-commissioned rank and those holding no rank or office at all.

While shades of militarism may still persist, much of the time in truly professional armed forces, informal, interpersonal relationships[33] exist between persons in a team working closely together, to good effect. Military cultures, conversely, maybe too informal, rendering them less than effective. Workable informality of relationships is probably easier to achieve in voluntary armed forces. Conscripts tend to be deeply divided from cadre non-commissioned officers and career officers and healthy informal relationships are unlikely to be formed as part of the military culture.

An indication of the high quality of professionalism within armed forces is therefore the clarity of status, an understanding of the need for hierarchies and rank structures, the quality of formal relationships and equally importantly, the quality of informal relationships between real, live human beings working together in difficulty or danger or risk. An understanding of when one should act formally with a senior or subordinate person, or switch to informal means of communication and interpersonal action, often enables the whole military endeavour to have a greater likelihood of success.

The penultimate attributes of professional, effective and 'usable'[34] armed forces are those of

- Leadership,
- Ethos,
- Morale and
- Reputation.

These are variable behaviours, beliefs and understandings, sometimes dependent upon and at other times independent of, the previous ten factors, circumstances and dynamics in this chapter. In civilian organisations typically 'up to 85 per cent of a corporation's value is based on intangible assets'.[35] What human, intangible factors should predominate in effective armed forces?

Numerous, mainly inconclusive works have been written and opinions expressed about leadership. Two significant factors spring to mind. Effective leaders are successful. They bring high quality to their own performance and that of those they lead, collectively and individually. Essentially, leading is an inspirational activity, which by definition is spiritual in quality and performance.

Professional armed forces, as a qualitative indication of their standing, manifestly require to be commanded by officers (commissioned and non-commissioned) who are effective leaders. Motivation is a variable factor, based on volition and choice. Strong motivation, self-directed choice and intelligent obedience are likely features of professional, volunteer armed forces. A definition is appropriate.

> To add the many traditional definitions of what leaders do, practically, intellectually and psychologically, I believe we should now recognize that at any one moment or instance, when one or more physical or conceptual military activities are being enacted, alongside can be found this other, coincidental moral activity – leading. This is initiated and sustained by the leader. ...Leading is about the extraordinary quality of the actions, above the ordinary mechanics.[36]

Two things are happening coincidentally, one observed and maybe measurable, the other felt emotionally, and psychologically.

Morale is dependent not only on internal factors, such as listed in all the forgoing indicators of professional quality, but also significantly in terms of reputation among the civil population and other professions. Reputation 'that immortal part'[37] of a person or institution is of particular sensitivity in life and death occupations. A high reputation is dependent on high standards of internal leadership, morale and effective and faithful performance. All of these factors are dynamic and variable. Even modest failure can affect morale and reputation severely and undermine ethos. So the best guarantors and trustees of morale, ethos and reputation are leaders, both military and civilian holders of office.

The ethos of an institution is, again, an intangible, spiritual quality. Briefly, it is ethical culture. 'Ethics differ from morality in that conduct may be described as "moral" when it is maintained or observed as fact, but becomes "ethical" as it rises from fact to ideal.'[38] Ideals, values, beliefs, performance are all part of the ethos of armed forces. High quality manifest in these attributes, is an aspiration of professional persons who take their calling seriously.[39] 'Ideal excellence' and 'the "genius"of an institution' are other, dictionary meanings of ethos.

The morale of armed forces is closely linked to standards of leadership and the corporate confidence and commitment individual members feel, based on its ideals and ethos. Low morale is the consequence of fear of failure, or actual failure. It is moreover dependent on numerous situational factors, personal relationships and events. Some of these can be measured and an assessment made of their consequences for corporate morale[40] and commitment.

The final indicator of professional armed forces is the evaluation of defence output and the validity of the evaluation process. This is best expressed in the form of questions.

- Is the institution maintaining, improving or witnessing a decline in professional standards, generally and in particular?
- How valid and sensitive are the techniques of measuring and qualitatively assessing corporate professionalism?

Armed forces in many nations have means of reaching their own answers and judgement. Some have highly sophisticated and comprehensive measurement systems; others are complacent, and have little means of measuring professional standards; a third group do not contemplate or even recognise military shortcomings. Potential allies or antagonists often attempt assessments on each others armed forces as part of diplomacy negotiations or intelligence activities. The utility of answering these questions is obvious.

The final evaluation of what might be termed national 'defence output' is a matter for governments, advised by civilian and military officials. Discussion of this aspect, albeit closely linked with the proportion of defence spending against gross domestic product and professional performance, is beyond the scope of this chapter.

CONCLUSIONS

The only appropriate conclusions and summary to this chapter is to comment briefly on internal and external reflexivity. It will have struck the reader that all these indicators of professionalism are interconnected and interdependent, both directly and indirectly.

Nations get the armed forces they deserve, just as they do governments. Circumstances change, as do perceptions, often for the better. The perceptions of professionals and the public, based on retaining a firm connection with reality as it affects the objectives of professional practice, is crucial. Combat or operations other than war, including peace support, are the *raison d'être* of armed forces. Professional failure is therefore frequently dangerous, and dangers are not necessarily apparent before the event. Military success has to be and is the culminating point of all trust, trusteeship and professionalism within the profession of arms, as well as among those who direct and support their nations' armed forces.

REFERENCES

1. Samuel P. Huntington, *The Soldier and the State* (Cambridge, MA: Belknap Press, Harvard University Press 1957) p.3.
2. The distinction is drawn here between organisations and institutions. The latter normally implies the characteristics of professional responsibility, regulation and accountability, with ethical codes and behaviour placing quality of service ahead of financial consideration, as well as requirements for the institution's self-perpetuation and self-rejuvenation.
3. Jacques van Doorn, *The Soldier and Social Change* (Beverley Hills, CA: Sage Publications 1975) p.29.
4. Article of Annex 15 to the 1907 Hague Convention IV.
5. In the mid-1970s there was a popular TV programme about policing, by violent means, of the same name.

6. 'Radical-modernity' or 'high-modernity', are terms probably more appropriate and helpful for practical purposes. 'Post-modern' was originally an artistic and cultural label, now rather dated in 2003. See Anthony Giddens, *The Consequences of Modernity* (Cambridge: Polity Press 1990).

7. General Sir John Hackett chose this phrase as the title for his celebrated and seminal book, *The Profession of Arms* (London: Sidgwick & Jackson 1983).

8. Svante Beckman, 'Professionalization: Borderline Authority and Autonomy in Work' in Michael Burrage and Rolf Torstendahl (eds.), *Professions in Theory and History* (London: Sage Publications 1990) p.117.

9. Dietrich Rueschemeyer, 'Professional Autonomy and the Social Control of Expertise' in Robert Dingwall and Philip Lewis (eds), *The Sociology of the Professions* (London: Macmillan Press 1983) p.41. This accords with 'personality' one of three sources of power, the others being 'property' and 'organization' in J.K. Galbraith, *The Anatomy of Power* (London: Corgi Books 1983).

10. Hannes Siegrist, 'Professionalization as a Process: Patterns, Progression and Discontinuity' in Burrage and Torstendahl (note 8) p.205.

11. Celia Davies, 'Professionals in Bureaucracies: the Conflict Thesis Revisited' in Robert Dingwall and Philip Lewis (eds.), *The Sociology of the Professions* (London: Macmillan Press 1983) pp.177–94.

12. See Rueschemayer (note 9) pp.38–58.

13. Andrew Dunsire, 'The Concept of Trust' in Rosalind Thomas, *Teaching Ethics, Volume I Governments Ethics* (Cambridge: Centre for Business and Public Service Ethics 1989) pp.336–77.

14. The permanence of civil servants, that is freedom from political motives in their appointment, is a measure of the term 'professional' in one sense – distinct from 'political'. Some nations, notably the USA, include numerous political appointees in their executive, legally and by custom.

15. Hew Strachan's book, *The Politics of the British Army* (Oxford: Clarendon Press 1997) is a historical view of politico-military tensions and activities in Britain.

16. *British Defence Doctrine* (MoD: Joint Warfare Publication, JWP 0-01, 1996) pp.3–5 to 3–14.

17. Some difficulties concerning national-military identities are explained in Patrick Mileham, 'But will they fight and will they die?', *International Affairs*, Vol.77, No.3 (2001) pp.621–9.

18. The British Army in Northern Ireland is firmly ' in support' of the civil police, many of whose members now act in the gendarmerie role when necessary. In the early 1970s, the Army de facto were in the front line, literally, when civil policing failed.

19. See Giddens (note 6) and Honora O'Neill's, *Reith Lectures,* BBC, Radio 4, April–May 2002, on the subject of 'trust'.

20. Measuring value placed on institutions of democratic states in Western Europe, parliaments, judiciary, police, press, education, armed forces etc. Quoted in *Daily Telegraph*, 23 September 1991.

21. Tom W. Smith and Lars Jarkko, *National Pride: Cross-National Analysis*, Report No.19, May 1998, University of Chicago, National Opinion Research Center. Willingness to defend the nation is an often surveyed question.

22. David Tickner, unpublished survey of recruits in the British Army in 1994.

23. Pilot, engineering, medical, legal qualifications, for example.

24. Say, Major 'in Command' and 'passed Staff College'.

25. Stanislav Andreski, *Military Organization and Society* (London: Routledge and Kegan Paul, 1954) pp.127–8.

26. Including command, control, communications, computers, intelligence and interoperability (C⁴I²) in NATO terminology.

27. Stockholm International Peace Research Institute; International Institute for

Strategic Studies, London: Royal United Services Institute for Defence Studies, London.

28. Michael Codner, 'The RUSI Index of Martial Potency' in *International Security Review* (London: Royal United Services Institute 2000) p.305.

29. Army Doctrine Publication, Volume 5, *Soldiering the Military Covenant*, Ministry of Defence, Army code no 71642, February 2000, p.1-1. An interesting point was raised by USAF Colonel Robert Marr, posing the question as to whether he could have ordered the pilots of the only fighters available in the air nearby on the morning of 11 September 2001 (four, all unarmed) to fly into and bring down the airliners heading for New York and Washington – thus proving unlimited liability of combatant servicemen. Reported in *Daily Telegraph*, 4 September 2002.

30. Max Weber, *Wirtschaft und Gesellschaft* (Köln 1964) p.866.

31. Colonel R.H.W. Crawford, 'Officer Training' in *British Army Review*, No. 81 (1985).

32. Tickner (note 22).

33. See Charles Kirke's four socio-anthropological 'structures' in 'A Model for the Analysis of Fighting Spirit in the British Army', in Hew Strachan (ed.) *The British Army: Manpower and Society into the Twenty-first Century* (London and Portland, OR: Frank Cass 2000) pp.227–41; also this volume.

34. A significantly expressed requirement in the Introduction to the publication of Britain's *Strategic Defence Review, Modern Forces for the Modern World,* 1998, p.1. The contrast is with 'Forces in being', i.e. for deterrence and display only, not for 'use'.

35. David Norton's Foreword to Brian Becker, Mark A. Huslid and David Ulrich, *The HR Scorecard: Linking People, Strategy and Performance* (Boston, MA: Harvard Business School Press 2001) p.ix.

36. Patrick Mileham, 'The Failure of Military Leadership', in Patrick Mileham and Lee Willett (eds.), *Ethical Dilemmas of Military Interventions* (London: Royal Institute of International Affairs 1999) p.50.

37. William Shakespeare, *Othello,* Act II scene 3; line 266.

38. Rosamund Thomas, *The British Philosophy of Administration* (Cambridge: Centre for Business and Public Sector Ethics 1989) p.141.

39. Vocation, in the traditional sense of the word, may be already lost in post-modern lexicography.

40. See Patrick Mileham, 'Morale in Armed Forces', *RUSI Journal*, Vol.147, No.2 (April 2001) pp.46–53.

6

The Changing Macro-Environment (1979–2001): The Implications for the Recruitment of Graduates into the British Armed Forces

BRIAN HOWIESON
University of Edinburgh
and
HOWARD KAHN
Heriot-Watt University, Edinburgh

For the British Armed Forces, the pace of change since 1979 has been extraordinary, with simultaneous changes in both international security and domestic society.[1]

This 'pace of change' has been caused by key macro-environmental factors: major political change in the UK; dynamic upheaval in the former Soviet Union; globalisation of the economy; vast and unprecedented advances in technology; changes in labour markets, employment patterns and organisational structures; and significant changes in societal values. Since 1979, these macro-environmental factors have had numerous effects, not least on the interaction between the British Armed Forces (BAFs) and modern society. This chapter reports on a study undertaken to determine the attitudes of graduates to a career in the British Armed Forces.

THE BRITISH ARMED FORCES (BAFS) AND PRESENT-DAY SOCIETY

Over the last 20 years, the BAFs have attempted to modernise their professional practices; however, little attempt has been made (during this period) to address the cultural and moral changes taking place in society. As late as 1998, many senior Army Officers were still arguing that, 'It was the duty of the Army to remain half a generation behind civilian society.'[2]

This modernisation of the professional practices (of the BAFs) without a corresponding appreciation of the changes in the cultural, morals and values of the youth in the UK, created internal strains – to say nothing of the external contradictions – with the society from which both its officers, non-commissioned officers and other ranks were drawn. The 'institution' that had sailed, largely untouched, through the so-called 'revolutionary' 1960s, found itself shaken by an utterly different ethos: a new generation and a different type of recruit was joining.

Specifically, many new recruits had little understanding or sympathy for tradition and convention. Indeed, the clash between the aspirations, attitudes and expectations of these new recruits, with the old ethos of collective loyalty, was unsettling: 'Many young officers remarked that loyalty to the BAFs could no longer be just a one-way process.'[3]

Significant traits (characteristic of these officers) during the 1980s and 1990s were inexperience, a different view of reality and what can only be described as ruthless ambition. Observation indicated that a large number of new recruits demonstrated a longing for instant professionalism and gratification; indeed, the video-generation had a relatively short attention span and wanted to 'fast forward' through anything that they perceived as boring.

Moreover, as discipline in schools declined and the rate of divorce and lenient parenting increased, young men and women appeared to be less impressed by authority than in previous generations.[4] Indeed, Canadian research has suggested that potential recruits perceive military organisations as bureaucratic institutions that are authoritarian and coercive.[5]

These facts have been echoed elsewhere: American research argues that Armed Forces (AFs) must now not only provide for the everyday sustenance needs of individuals, but must also liberate their creative drives, which can be difficult in rigid command and control, bureaucratic structures.[6]

The popular press, meanwhile, encouraged people to believe that they could be masters of their own destiny, and that if anything went wrong, somebody else had to be responsible. Indeed, the tabloid press was guilty of an ambivalent attitude to the BAFs, 'They liked to support the nation's service personnel; however, they were quick to fault and denigrate the military institution and its leaders.'[7]

In parallel, public attitudes towards the BAFs altered significantly as individual perception of the BAFs were gained from media portrayals.[8] The media tended to focus on the 'peculiarities' of the BAFs, rather than presenting a more accurate picture: 'Today, the misconception that the British Army is a refuge for dullards still exists.[9]

Critically, the BAFs had to recruit and train for one of the most unpredictable and dangerous occupations of all, that of active service. Even the chaos of battle was no longer regarded as an excuse: tragic accidents, such as fatalities and wounds from combat, were no longer accepted as acts of God. As the Ministry of Defence (MoD) discovered to its cost, society had entered not just a secular age, but also a litigious one. If the experiences of the Police Force are to be followed as an example, the BAFs will see a growing rate of legal cases dealing with thwarted promotions, hurt feelings or the effects of stress.[10]

The BAFs, because of their origins and as a result of their hierarchical structure, long believed in respect for rank as an essential prerequisite to discipline; however, they have been exasperated by the idea that they should have to reduce their own exacting standards of behaviour to the declining levels of morality in society as a whole. In the UK today there is little patience or sympathy for past practices among the younger generation. Young men and women may have gained their values, morals and ethics from current political and social icons, and values based on deference, hierarchy and collective loyalty are seen as ridiculously old-fashioned. What distinguishes today's youth from their predecessors is the rejection of non-conformity, coupled with a new determination to become adult as quickly as possible.[11]

Furthermore, as organisational structures in the civil world flattened (and developed a previously unimaginable informality) in a rapidly changing workplace, 'institutionalised society', predominant in the greater part of the twentieth century, was coming to an end[12]. At the start of the twentieth century, the German sociologist, Max Weber,[13] offered a vision of a society dominated by large bureaucratic organisations that had unique characteristics: pyramidal in shape; a clear hierarchy of positions; a well-developed system of codified rules and norms; and appointment and promotion according to open and testable procedures. As organisations have now moved towards the empowerment of the worker (with more emphasis placed on self-managing teams and autonomous work units), hierarchical command, a defining characteristic of the military, may now be contradictory to the attitudes and values of the youth of the UK today.

In addition, since 1979 Western society has slowly moved from an industrial economy to a service economy with a greater emphasis by employees (in both the private and public sector) on flexible working conditions, bonus and remuneration packages, performance-related pay and on obtaining immediate results from one's efforts. In tandem, there has been an increasing importance on theoretical knowledge, an increased desire for leisure pursuits and economic well-being, and a significant growth in 'humanistic' values, in contrast to the values of the Calvinistic work ethic.

In contrast to the civilian world, the BAFs maintain a 24-hour a day claim on the serviceman/woman; as one soldier said (of his decision to leave the Army): 'I would prefer to get myself a nice little 9 to 5 job, come home and my mum is making dinner, and after work my time is my own.'[14]

Finally, increased education and access to information, on a scale unimaginable even a quarter of a century ago, has made the youth (in society today) more knowledgeable, 'savvy' and demanding than those of yesterday. The recruit of tomorrow will have grown up in a society where information is freely available and will express considerable independent thought. Tomorrow's potential recruits will be more aware of their rights, disregarding of their responsibilities and reluctant to accept responsibility for their actions. Respect will need to be earned and blind acceptance of authority will not occur: 'The recruit of tomorrow will be sceptical, will doubt and question everything and will need convincing.'[15]

Graduates: The Military Personnel of the Future

In 1984, 60 per cent of the entrants to work had some qualification; by 1995, this figure had increased to around 80 per cent, and in the 25–34 year group, this figure was 86 per cent. More students are now staying on at school after the age of 16: in 1984, less than half of the 16-year-olds stayed on at school, and by 1995 this figure had increased to 75 per cent. At the same time, enrolments in colleges and universities increased from 1.5 million in 1981 to 2.6 million in 1995 and recent press reports suggest that the UK government aims to increase this proportion to 50 per cent of school leavers within the next 8 years.[16]

Therefore, trends suggest that people will spend longer in continuous full-time education and will, on average, be better qualified.

The average age of the new employee is likely to be older; the consequence is that individuals will have higher expectations about their careers, be more mature and socially less impressionable. What has conditioned the minds of young people in recent times has been the spectre of unemployment; it has shown them that the world is not just going to give them a desirable career, but they will have to earn a career via personal development, academic training and skills acquisition.

Therefore, the overall picture of the UK is of an increasingly well-qualified population: the main consequence of the increased participation in tertiary education is that the supply of bright 18-year-old school leavers will diminish markedly and it is, therefore, likely that all but the few (less-skilled) BAFs personnel will be graduates. As a consequence, the BAFs will finally have to surrender its desire and practice of attracting bright, adventurous school leavers and training them (as if still in boarding school) to become officers. It will have to recruit its officers and other ranks from the population of graduates.

Graduates: a Key Resource to the Long-Term Competitiveness of the British Armed Forces

The graduate labour market is the interface where employers seek the highly educated young men and women that they need in order to ensure their long-term competitiveness; it is from that an increasing proportion of the brightest in society enters the world of work. In the last 20 years, the demand for graduates among traditional recruiters in the UK has grown steadily, in parallel with economic expansion. Competition, technology and rising consumer power are changing the way work is defined, the skills and jobs needed, and where they are located. Traditional boundaries are fast disappearing as organisations move into new activities and cross-sector alliances, and seek to resource operations internationally. As a result, the staffing profile sought is shifting to higher level occupations and to higher skill profiles within these occupations.

The fast pace of organisational change and the increasing use of Information Technology/Information Systems (IT/IS) has meant that organisations now and in the future will rely more on personnel with the skills which can only be obtained via tertiary education. Over the longer term, the number of managerial, professional, technical and high level jobs will grow: this will lead to a rising demand for graduates.[17] The nature of jobs treated as 'graduate jobs' will continue to change, with many technical and associate professional jobs being filled by graduates. Jobs in general are becoming more complex and demanding, and for the BAFs the critical challenge will be to recruit graduates with the right skills and competencies.

Indeed, one of the effects of the period of international and domestic turbulence since 1979 is that the 'people dimension' of the BAFs has become a central problem in defence debates: 'The pool of young people available for recruitment and manning will be the single most critical element to shape the AFs of the 21st century.'[18] The graduates most in demand will be those who combine intellectual and personal attributes in areas such as team working, motivation and communication along with the ability to continue learning. Therefore, there is expected to be excess demand for the 'best' graduates. As a consequence, these graduates, with superior qualifications from the best universities, are now an increasingly powerful and demanding group; they are extremely confident and are forcing employers to fulfil their (the graduates') expectations or risk losing the war for talent. As a result, BAFs will need to recognise the diverse nature of graduate supply and acknowledge the competitive environment of the graduate labour market.

THE CHANGING MACRO-ENVIRONMENT

The macro-environment (in which the BAFs operate) is in a state of constant flux and numerous external factors have re-shaped both society and the military environment, including political/legal, demographic/economic, socio-cultural and technological factors.

The Political/Legal Environment

Despite questioning the central role of the state, it is important to note the significant impact of the legal context on the UK military. The focus from supra-national agencies has been on issues such as health and safety at work and employment law.[19] Indeed, European Union legislation is phasing in significant rules to protect the rights and well-being of employees.[20] In simple terms, this legislation reflects the growth of individualism within society, and aims to balance the rights of the individual with that of collective responsibility.

The Demographic/Economic Environment

Over the last 30 years, the demographic profile of the UK has undergone significant change in terms of age, gender and ethnic composition. The UK has had a growing and ageing population: this trend seems set to continue with the percentage of the population aged over 65 years rising from 11.7 per cent in 1960 to a projected 19.3 per cent in 2020.[21] Increasing numbers of 16–24 year olds are likely to be available up to 2012, but this will be followed by a decline. The number of those in the UK population aged between 16 and 39 will be 19.7 million in 2001: in 2021 it will be 18.6 million and in 2031, 17.4 million.

Importantly, these trends indicate that there will be greater competition for workers in the younger age groups. By 2006, it is estimated that the number of over 35s in the labour force will have risen by 2.3 million since 1999, while there will be 1.1 million fewer under 35s. Additionally, the participation in the labour force by females has been increasing steadily since the 1960s and this trend is set to continue: by 2006 women will account for almost half the total labour force (in 1971 women represented only 37 per cent).

Furthermore, the trend of employment patterns has been towards a shorter lifespan of employment and less stable patterns of employment. Short term and temporary contracts are becoming increasingly common, enabling organisations to adapt and change (in response to unexpected needs and trends) without incurring undue financial penalties. These largely commercially-driven changes are expected to continue and will bring with them lasting changes to the expectations associated with employment. Organisations may not be able to offer a job for life anymore; however, for many young people entering the job market, this is no great deprivation. This cohort has seized their employers' demands for employability, and has used this requirement to their own advantage, forcing organisations to compete for the most employable skills.[22] Thus, in the future, people will probably work in a range of sectors within their working lives, learning a variety of different skills.

The Socio-Cultural Environment

Societal trends in the UK have generally reflected those of the Western world. Greater global and local mobility has been combined with an ever-increasing emphasis on human rights, individual rights, equal opportunities and diversity, and intra-national, multi-ethnic issues, with less emphasis on group identity and responsibility including the family, national and other groupings.

One of the most significant current trends in the UK, is the decreasing respect for social institutions. For example, the government, church, monarchy, judiciary and the police have all been subjected to greater public scrutiny, and in many cases this is due to a failure by these institutions to adapt to the rapidly changing needs of ever more demanding citizens:

> The institutions that people believed in – the government, legal system, police, the family, the housing market and even the quality of British food – are no longer reliable.[23]

However, the social institution that has disintegrated most

rapidly is the traditional family. In recent years, there has been a major increase in the rate of divorce, single parent households, cohabiting couples and people living alone. The family structure has been weakened: there are now reduced role models or moral anchors for (bright) young people.[24] There are fewer marriages, people are marrying later in life and they are having fewer children. In addition, there has also been evidence of changes in household structures. Such changes include significant increases in the numbers of single-person households, dual-career couples and single parents.

More generally, there is an increased tolerance for diversity in private lifestyle and a reaction against any attempt to interfere with choice in such matters. Thus, British society is more 'permissive' and the emphasis is on individual rights rather than a responsibility towards the community or collective organisations such as the BAFs. Therefore, as many of the influences once supportive of positive attitudes towards authority have waned, the BAFs' core institutional values of duty, self-discipline, self-sacrifice, respect and concern for others, are no longer rooted firmly in wider society. Legislation and public attitudes are now much more based on a 'rights-based' ethic than two decades ago. Whatever their basis (sexual, political, religious preference, ethnicity and so on), groups now express rights of freedom and self-expression, and ask for all barriers to such expression to be removed.

Furthermore, for both policy-makers and employers, questions of work-life balance are becoming permanent agenda items. Whether for reasons of economic or labour market efficiency or out of a sense of corporate responsibility, employment policies that facilitate the balance between work and life demands are currently under scrutiny.[25]

A mix of coercion, a tight labour market and shifts in the employment relationship have driven employment behaviour. Therefore, the concerns of educated young people are not solely related to their identity as consumers; apart from their rights as citizens and consumers, they are also interested in achieving that balance in their lives that they believe to be appropriate. This especially relates to an appropriate division between working life, social life and responsibilities.[26]

There is now a trend towards viewing work as just one component of life rather than the 'be-all and end-all'. The demands of permanent employment and steady progression up a career ladder can be perceived as overly restrictive and some organisations now attempt to counter this by offering career-break schemes and sabbaticals, which allow individuals to develop by undertaking some non-work activity, such as travel. At the same time, the assumptions frequently underpinning traditional careers are those of full-time employment (often involving long hours) and undertaken by one individual from a family unit (usually male): in the light of greater numbers of dual-career couples and the decline of the nuclear family, these fundamental assumptions become increasingly untenable.

Finally, many of the discussions around the transformation of work in the UK have centred on the concept of the Psychological Contract (PC), referring to the implicit mutual expectations of employer and employee.[27] The PC is an unwritten set of expectations operating (at all times) between every member of an organisation and the various managers in that organisation.[28]

Many of the changes from the traditional to the new forms of employment have been typified by a transition from a long-term 'relational' PC (the promise of job security in exchange for loyalty to the organisation) towards a more short-term 'transactional' contract based on a more explicit negotiation between the individual and organisation concerning what each side expects to give and receive in return. The exact relationship between the BAFs and the PC in a military setting is not yet fully understood, and its renewal, in a more subtle form, is not yet underway.[29]

The Technological Environment

Every indication is that the 'Information Revolution' will continue to gather pace and affect almost every area of daily life. Electronic business processes will develop rapidly, affecting strategies and structures, politics, culture, regulations and finance. Importantly, the increased use of digital technology for the integration and handling of information will impinge upon employment and careers.[30]

The nature of the task has been transformed and thus, instead of carrying out work involving monotonous and repetitive tasks, technology has resulted in growing abstraction and complexity in data interpretation and processes at work.

Moreover, management styles are changing to reflect a new emphasis on creativity rather than a task culture approach: these management styles will place greater emphasis on knowledge and expert power rather than position and status power.

Key areas

Key areas from this macro-environmental change – since 1979 – are as follows:

- The supply of bright, adventurous 18-year-olds will diminish markedly (due to changing demographics).
- As more young people undertake further and higher education, the average age of all recruits to the BAFs will be older; the consequence is that these individuals will have higher expectations about their careers, will be more mature, and socially less impressionable.
- Due to the fast pace of organisational change and the increasing use of IT/IS, the way work is defined and the skills and jobs needed have all changed. As a result, the staffing profiles sought have shifted to higher-level occupations and higher skill profiles within these occupations. Therefore, all major employers now have an increasing demand for personnel with the skills only obtained via tertiary education.
- The requirement for employees to change as the organisation changes in response to its environment will be paramount. Emphasis will be on a high degree of data processing capacity; the ability to extract information from this data; the ability to communicate it clearly and quickly; and the capacity to make important decisions, very rapidly, under pressure. As a consequence, the personnel most in demand will be those with the ability to learn new skills rather than those possessing existing skills. Generic skills and values are becoming highly prized; these include business awareness, commercial,

interpersonal and communication skills.
- The competition for the 'best' graduates will become intense. For the BAFs, the critical challenge will be to recruit graduates in a very competitive and shrinking labour market.
- The notion of the Psychological Contract between employer and employee is important in defining the contemporary career: graduates now want more responsibility for their own careers as employers take less; they are looking for more varied types of career that meet their own individual needs and aspirations. Graduates want to be valued and challenged intellectually, to have personal control over their careers, and they seek menu 'benefits' packages.
- As the structure of the workforce changes, with greater emphasis on a task culture rather than mechanistic structures, intelligent and flexible graduates will not be content to work in traditional military hierarchies.

In summary, the last 20 years have proved to be a period of radical change for advanced societies, including their military institutions. The main problem is that the changes stemming from the external strategic environment and the domestic social structure are not occurring sequentially but simultaneously.

THE FUTURE SITUATION

The Increasing Demand for Graduates by the British Armed Forces

Looking ahead, the econometric forecasts in the UK and most Organisation for Economic Cooperation and Development economies point to a continuing growth in the demand for personnel with high-level skills. The latest UK econometric forecasts suggest that the numbers employed in managerial, professional and high level IT/IS occupations will grow by 20 per cent to the year 2010: this will lead to a rising demand for graduates by all employers. The nature of jobs treated as 'graduate jobs' will continue to change, with many technical and associate professional jobs, which were previously filled by personnel with lower academic qualifications (e.g. Higher

National Diploma/Higher National Certificate [HND/HNC])
being filled by graduates.

These occupational changes will mean that the BAFs will
require an intelligent and flexible workforce that can respond
to what are often ill defined, yet significant, challenges. This
fact is also being echoed by all major employers in the UK. The
main driver for the competitive advantage of the BAFs will be
the intellect, skills and learning capability of its personnel.
Therefore, recruiting the 'best' will be a major challenge for the
BAFs and will feature high on the corporate agenda.

The Supply of Graduates to the British Armed Forces (BAFs)

The supply side of the labour market equation is equally
important: each year, major recruiters in the UK seek more
than 20,000 new graduates to enter management and
professional training schemes.

Although there will be an increase in the numbers of
qualified graduates, there will also be an increase in their
diversity: in 1995/96, half of the 1st year undergraduates were
over 21 years of age and more than half of this cohort were
female. In 1996, 57 per cent of ethnic minorities (aged 16–24)
were in further or higher education compared with 41 per
cent of whites.

As a result, the supply of graduates available to the BAFs
will become more diverse in age, socio-economic background,
knowledge and skills. Therefore, the BAFs will need to
address the following issues: are there potential sources of
new recruits that have not been considered hitherto; what
features of employment in the BAFs are most likely to attract
applicants from these new sources; and what is the best way
to market military careers to appeal to these new sources?

Key Areas

Key areas from this analysis are as follows:

- The Next Generation of Service Officer (NGSO) will be
  recruited from a diverse and fragmented society.
- IT (as a channel of learning) will become dominant,
  resulting in complete and universal ease with this

medium by the majority of the youth in society. As a result, graduates will have complete familiarity with IT and the Internet and will enjoy the freedom of being able to 'windowshop' online as part of their job search process. Therefore, as technology changes not only the traditional nature of work, but also the entire recruitment landscape, the 'brand' of the BAFs will become increasingly instrumental in attracting and recruiting the supply of graduates from the education system.

• The BAFs will have to manage the expectations and understand the needs of the 'new wave' graduates at a time when the profile of the typical graduate has been eroded and is now difficult to define.

QUANTITATIVE AND QUALITATIVE INVESTIGATION

In light of the significant macro-environmental change (since 1979), that has impacted on the BAFs and present-day society, the authors sought to illicit quantitative and qualitative information from the 18–25 age group cohort in UK society on several dimensions:

• The career of the year 2000 and beyond (including financial and non-financial issues).
• Comparison with military and civilian employment.
• Attitudes and perceptions towards a military career.

In addition, the authors interviewed specialist recruitment staff in the civil sector of the UK labour market and recruiting personnel in the BAFs/MoD, in order to obtain 'recruiters' views of the above issues.

Quantitative Research

• Questionnaire One: 300 questionnaires were distributed among Officer Cadets (OCs) at the Britannia Royal Naval College Dartmouth, the Royal Military Academy Sandhurst and the Royal Air Force College Cranwell to determine what were the major dimensions of the career today and what relative weightings were placed (by the OCs) on the various aspects of the career in 2000 and beyond.

- Questionnaire Two: 500 questionnaires were distributed among 1st year undergraduates at the University of Edinburgh and Heriot-Watt University, Edinburgh, to determine their perceptions and attitudes towards a career in the BAFs.[31]

Qualitative Research

- Interviews with recruitment specialists from private and public sector employers in the UK to determine their approach and strategy towards graduate recruitment.
- Interviews with personnel from the BAFs/MoD to determine their approach and strategy towards graduate recruitment.

RESULTS FROM THE QUANTITATIVE AND QUALITATIVE INVESTIGATION

Questionnaire One[32]

This questionnaire produced interesting results: Officer Cadets desire financial benefits and remuneration packages in line with their civilian counterparts (50 per cent of the OCs wanted performance-related pay, 90 per cent of the OCs wanted their remuneration to reflect their knowledge-base and skills set rather than the current 'pay-per-rank-system', 65 per cent of OCs wanted an annual bonus system to be introduced, and the majority of the OCs thought that pay differences between certain branches caused divisiveness).

The OCs desire flexible careers and short-term contracts and they will not commit to a long-service-in-one-organisation career (almost 80 per cent of the OCs saw their careers in the British Military to be short term in nature and saw the BAFs to be a 'stepping-stone' to something else).

No fewer than 93 per cent of the respondents saw it as important that the BAFs are a forward-looking and dynamic organisation and are very uninspired with outdated modes of working; and 60 per cent of the OCs desired a personal choice in employment location and a greater say in career decisions and postings.

In addition, the OCs were asked to rate the top six dimensions of their career. The results are shown in Table 1:

TABLE 1
THE TOP 6 DIMENSIONS OF THE CAREER FOR OCs

RANK	DIMENSION
1st	Job satisfaction
2nd	Varied work
3rd	Good salary
4th	Adventure/travel
5th	Intellectual challenge
6th	Friendly colleagues

Finally, two qualitative questions were asked of the OCs: what were their concerns with future military careers and how did their military employment compare with civilian life?

Concerns with future military careers included injury during training; death in combat; becoming institutionalised in the military establishment; family and relationship issues; discipline and bullying, length of commitment to the BAFs; and further defence cuts.

Comparisons with civilian employment revealed that the OCs saw a relative imbalance with their salary compared to those personnel in civilian life: in other words, money was deemed important when the nature of a military career was compared directly with other civilian and less dangerous professions.

Questionnaire Two[33]

The 1st year undergraduates were asked to rate the five most important dimensions of their career. The results are shown in Table 2:

TABLE 2
THE TOP 5 DIMENSIONS OF THE CAREER FOR UNDERGRADUATES

RANK	DIMENSION
1st	Salary
2nd	Intellectual challenge
3rd	Working conditions
4th	Friendly colleagues
5th	Varied work

TABLE 3
UNDERGRADUATES' REASONS FOR NOT CHOOSING A
MILITARY CAREER

RANK	DIMENSION
1st	Discipline
2nd	Career opportunities are better elsewhere
3rd	Not a useful career
4th	Anti-military
5th	Can be sent abroad at short notice

The undergraduates were then asked why they would not choose a military career. The results are shown in Table 3.

In addition, the questionnaire revealed that the majority of the sample would not consider a career that is dangerous and that involved risk to themselves, and did not like imposed discipline and hierarchical work structures. A total of 88 per cent of the sample sought personal control over their career moves; the majority wanted to work for employers who had contemporary equal opportunities policies; and a majority of respondents did not view the UK military as representative of society.

Again, two qualitative questions were asked of the undergraduates: 'what would scare you about a military career?' and, 'what would the BAFs have to do for you to consider a career with them?'

A selection of responses to these 2 questions is shown in Appendices A and B of this chapter.

Interviews with Graduate Recruiters in the Civil Sector of the Labour Market[34]

The major themes from the interviews were as follows: graduate recruitment is taken extremely seriously and private sector employers recognise that the competitive advantage of their organisations rests with graduate skills. The private sector respondents have specific graduate recruitment strategies that include proactive strategies for monitoring competition in the graduate labour market; competitor monitoring is limited in the public sector.

All private sector organisations recognise the forces of demand and supply: the skills and qualifications most in demand are remunerated above the going rate.

All employers now recognise that 'ownership' of the career has shifted from the employer to the employee; and all employers recognise that salary, training and development, recognition, intellectual challenge, a balance between work and personal life, and choice in career decisions are important to graduates.

Interviews with Recruiting Personnel in the BAFs/MoD[35]

The major themes from the interviews were as follows:

On the demand side, no accurate or consistent definition exists for the officer in a changing society and changing BAFs; many personnel had difficulty defining the profile and job specification for the officer in 2000 and beyond. No thought has been given to job design and analysis of the knowledge, skills and abilities (KSAs) required for the Next Generation of Service Officer (NGSO), although the Army is conducting specific graduate recruiting seminars, graduate recruitment strategies are limited. Personnel Strategy Guideline 4 (Recruiting Policy) from the Armed Forces Overarching Personnel Strategy is dangerously limited, and does not indicate how the Ministry of Defence and the three single Services are going to develop and sustain competitive advantage in the graduate labour market.

On the supply side, no thought has been given to the career that the NGSO wants or expects; no thought has been given to managing the expectations and understanding the needs of the new wave of graduates. The concept of competitive position, competitive environment and competitive advantage in the labour market is not understood and competitor intelligence and monitoring is not being undertaken. No thought has been given to the likely changes in traditional sources of personnel and into probable new sources. The BAFs are falling behind other recruiters in terms of the sophistication, immediacy and personalised character of their recruitment, while significant emphasis on recruitment to the BAFs is still via Armed Forces Careers Offices.

CONCLUSIONS FROM QUANTITATIVE AND
QUALITATIVE RESEARCH

Questionnaire One

The main conclusions are as follows:

- The management of careers and rewards must be increasingly flexible and take account of a wider variety of individual differences.
- OCs do not regard qualifications acquisition to merit promotion, as long as they are remunerated for these skills/qualifications in line with their civilian counterparts.
- The pay/rank remuneration tool must become more flexible: the scarcity of some skills will become such that their 'owners' will have to be differentially rewarded if they are to be recruited (and retained!).
- OCs (who are self-selecting and who might be expected to have positive attitudes to the BAFs) have expressed concern about the current BAFs career and its associated dimensions. Thus, the career that the BAFs presently offer, and its key components, be they financial or non-financial, may be acting as a barrier to potential recruitment to the UK military.

Questionnaire Two

The main conclusions are as follows:

- Salary is now a very important criterion in the occupational choice of undergraduates.
- Most undergraduates do not desire dangerous occupations.
- The BAFs are not perceived to offer a contemporary career.
- Undergraduates have very poor perceptions of and attitudes towards the BAFs.

Interviews with Graduate Recruiters in the Civil Sector of the Labour Market

The main conclusion is as follows:

- Private sector recruiters are taking graduate recruitment very seriously and recognise the intense competition which exits within this sector in the UK labour market.

Interviews with Recruiting Personnel in the BAFs/MoD

The main conclusion is as follows:

- There is a lack of understanding by BAFs/MoD recruiting personnel of the importance of the development and sustainability of competitive advantage in the graduate labour market, and specific graduate recruitment strategies are limited.

CONCLUSIONS

The following general conclusions are drawn. The demand for graduates by the BAFs will increase significantly and although the number of graduates in the UK will increase (as a proportion of the nation's youth), the absolute supply of graduates to the BAFs will decrease due to: demographic changes in the labour pool; the perceived failure of the BAFs to offer a contemporary career; poor perceptions of and attitudes to the UK military by undergraduates; aggressive competition in the graduate labour market; and no graduate recruitment strategy by the BAFs.

This imbalance of demand and supply will directly affect the compelilive advantage of the BAFs.

RECOMMENDATIONS

The Career

First, a complete review of the career that the BAFs currently offer should be undertaken immediately. This review must focus on the expectations and demands of the best graduates, who have the skills most in demand. In addition, this review must focus on all dimensions of the contemporary career, including financial and non-financial considerations.

Second, that an investigation should be undertaken into the contemporary desired psychological contract of graduates, to determine, in greater detail, why they do not wish to undertake military careers.

Finally, that a further investigation be carried out to examine the costs and benefits of de-coupling reward for rank and replacing it with reward for skills/qualifications.

Perceptions and Attitudes towards the BAFs

Attitudes impact upon the behaviour of individuals; therefore, the study of public attitudes toward the BAFs is vital. Figure 1.0 shows a simplified approach to attitudes and their impact on behaviour (arrows indicate proposed causal relationships). To date, very little work has been carried out with respect to attitudes concerning the military; this must be rectified. As attitudes towards the BAFs seem to be negative, there is a pressing need for systematic psychological research into attitudes and attitude change to the BAFs.

Critically, given that most individuals rely on media reports to gain information about the BAFs, it is imperative that the filtering effects of the media are acknowledged. Precise empirical study of the attitudes toward the BAFs and investigations into the impact of theory-driven interventions may produce positive changes in the short or medium term.

FIGURE 1
A SIMPLIFIED APPROACH TO ATTITUDES AND BEHAVIOUR

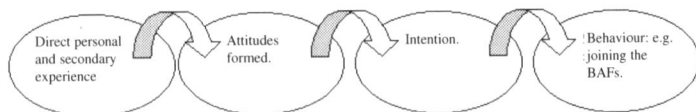

The Competitive Environment

First, that an investigation of the recruitment strategies of direct competitors to the BAFs in both the private and public sectors needs to be undertaken immediately. This investigation must consider their (competitors') strengths, weaknesses and future intentions, and how they (competitors') develop and sustain competitive advantage in the graduate labour market.

Second, that a strategic and analytical approach to continual and exhaustive competitor monitoring is also undertaken.

Finally, that an investigation of the recruitment strategies of allied armed forces should be undertaken to learn from their approaches and strategies towards graduate recruitment.

A Graduate Recruiting Strategy?

A recruitment strategy (that recognises graduates as the NGSO) should be implemented. This strategy must include: demand factors (research into establishing the profile and job description for the officer in a changing society and changing BAFs, for example, what KSAs will be required in light of the Defence Missions published in the Strategic Defence Review (SDR) and general strategic considerations of the extent to which existing recruiting strategies support these KSAs and how they could do so to better effect); and supply factors (research into the likely changes in traditional sources of personnel and into probable new sources).

AN ENDNOTE

Unless the BAFs take a conceptual leap into the future and then work back from this point, their recruiting policy will remain reactive and short term, rather than proactive and strategic in nature: projections 15 years ahead are hazardous and trends do not necessarily continue in the same direction. However, the context of the year 2015 (in line with SDR) is a necessary backdrop for the development of a recruitment strategy that will contribute to the development and sustainability of the competitive advantage of the BAFs in the graduate labour market.

APPENDIX A

Question: 'Would anything 'scare' you about joining the British Armed Forces?'
The clique attitude.
To kill.
Lots!!!!!!!!!
Ability to meet the standards required.
Getting hurt.
Physical training.
Armed combat.
Being on the front line.
Can be condemned by press for seemingly small crimes.
I don't feel that knowing that you are putting your life on the line would appeal to me.
Being shot in war.
I have friends who are in the Army: their experiences in Northern Ireland put me off.
The danger element of combat; away from friends and family; uncertainty.
The intense discipline.
Being sent into a war zone.
Male domination.
Being sent to war.
Uniformity.
The discipline – saluting and standing to attention.
Emphasis is on pain as character building.
Yes, hating it.
Getting up early and having to do lots of exercise.
Yes – potential death.
Attitudes of others in the Army.
Violence.
Accidents with equipment.
Bullying/having to kill people/de-humanisation.
Getting laid off early through cutbacks.
Using guns on people.
Getting shot.
Discrimination against homosexuals.
The physical activities/the way that they communicate with you – unnecessary shouting and lack of respect.
Being taken out of 'normal' society and being placed in such a structured hierarchical system with little room for self-expression.
Don't think that I would be able to have much of a life at the same time: constantly uprooted.
Not being able to leave – seems very inflexible.
The personal risk.
Uncertainty in day-to-day life; feeling inferior to male.

APPENDIX B

Question: 'What would the British Armed Forces (BAFs) have to do for you to make you think about a career with them?'

Nothing – not my kind of lifestyle.
The BAFs would have to make me more aware of what is actually involved in taking a career with them.
Provide the ability for self-expression without the loss of discipline and structure.
To be honest, the amount of discipline is a major factor.
Better knowledge of non-combative posts would be good.
There is nothing they could do to make me join: I am anti-war and do not feel I would have any place in the Armed Forces.
Gain a better reputation.
Introduce more women/take out all officers with power trips.
Realise that shouting is not a good way of motivating people.
Less discipline.
Nothing!
I am anti-military so I would not join the Forces, under any circumstances.
I would not consider it at all.
Show how it really is.
I would never join this kind of organisation.
Stop hurting people.
Make better and more detailed advertisements.
Highlight areas other than combat: business opportunities, career avenues, travel opportunities.
Not interested at all.
I would never consider a career in the BAFs.
Raise the salary.
Tell me that I would be treated with respect/not to have to undertake the gruelling physical activities.
Guarantee there would be no possibility of going into war/combat against any other country – no danger.
By force – no other way.
Make the job appear fun and open to everyone: to me it appears male dominated and does not seem like a good environment for families.
Make me more aware of the benefits/be more attractive to women/be more flexible.
Less structured and disciplined; more females; less competitiveness and more positive group dynamics; positive further career prospects e.g. after injury.
More relaxed, less strict, less physical – if taking a desk job, why do you have to do same physical training as front-line soldiers?
Ensure safety, more individualistic.
I could never join the Armed Forces – I could not be subjected to that level of discipline.
Less emphasis on being 'The Best.' Less fighting, aggression, militaristic attitudes.
Less discipline and the stigma that is attached to it that it is a 'man's job.'
Change their image: open to educated people.
Not brawn and no brain.
Make it more of an equal place in terms of class, gender and ethnic minority.
The idea of rolling in the mud shooting guns does just not appeal to me.
Become less closed in.
Change these things: no opportunity for unlimited wealth; not easy to have a family at the same time; work environment not aesthetically pleasing; too many 'petty' regulations that are unnecessary; and give me loads of money.
Simply not an intellectually stimulating job.
Become more approachable.   [continued overleaf]

135

Must reflect the society in which we live.
More training for skills.
Show that not everyone goes to war.   [Appendix B concluded]

REFERENCES

1. Christopher Dandeker and David Mason, 'Military Culture and Modern Society: Civil Military Tensions and the Management of Change', unpublished paper, 2000.
2. From Antony Beevor, 'The Army and Modern Society' in *The British Army, Manpower and Society into the Twenty-First Century* (London and Portland, OR: Frank Cass 2000).
3. Ibid.
4. For an expansion of this theme, see W.B. Howieson, *How has the Changing Macro-Environment Affected the Career Attitudes of Officer Cadets in the British Armed Forces?* MoD Paper: D/Pol Planning/12/4&12/4/38, 2000.
5. Explained at length in Hills, *The Military in a Changing Society: The Impact of Demographics on the Canadian Forces*, Canadian Department of National Defence, June 1997.
6. Center for Strategic and International Studies Report, Washington DC, February 2000.
7. M.A. Bagnall-Oakley, 'The Effects of Current Social Attitudes on Manning the Armed Forces', *British Army Review*, No.97 (April 1991) pp.34-43.
8. See *The Times*, 13 July 2001, 'Services reject "Blimp" Image': in this article, the current Chief of Defence Staff, Admiral Sir Michael Boyce vociferously defends the reputation of the BAFs against the views that all military officers are dinosaurs, elitist and 'blimps'.
9. Bagnall-Oakley (note 7).
10. For example see *The Times*, 27 May 2001; 'Police, paramedics and soldiers are claiming compensation for the hazards of their jobs'.
11. See 'Today's Teenagers, Tomorrow's Yuppies', an address by C. Restall of McCann Erickson.Advertising Ltd to the IBC Conference, 12 April 1988, p.1.
12. For example, see ASDA Graduate Recruitment Website, <www.asda.co.uk> 'We're not hung up on hierarchy...everyone who works for us is known by their first name, even Allan, our Chief Executive.'
13. Max Weber (1864–1920) was one of the most prolific and influential sociologists of the twentieth century. *From Max Weber: Essays in Sociology*, edited by Gerth, Mills and Turner(Oxford and New York: Oxford University Press 1973) includes a selection of his key papers that describe bureaucratic society.
14. Creative Advertising Research for the Army (job no: 2688), *The Research Business*, June 1989, pp.20–2.
15. Hills (note 5).
16. For example, see *The Times*, 17 November 2000, 'Universities get extra £1 Billion to fund expansion.'
17. In his article: 'Grim prophecy of war on every front', Michael Evans (*The Times*, 8 February 2001) describes the MoD's 'vision' of future wars. What is very clear from this vision, is the increasing complexity of the high technology weapons that will be used by the BAFs. This technology will require highly trained and educated operatives.
18. Quoted from an interview with the Chief of Defence Staff in *The Independent*, 20 December 2000, 'Army admits it is 8,000 personnel short.'
19. For example: The European Commission; The European Court of Human Rights; and The European Court of Justice.

20. The Armed Forces (AF) Discipline Act (2000) received Royal Assent on 25 May 2000. The system for administering discipline in the AFs is kept under review, with the principal vehicle for any legislative changes that may be necessary, being the five-yearly AFs' Acts. The AFs' Act (1996) made substantial changes, reinforcing the independence of courts martial to reflect the European Convention on Human Rights. The Human Rights Act (1998) incorporates certain provisions of the European Convention into domestic law and the main provisions of the Act came into effect on 2 October 2000. The MoD has used this as a framework for a further review of the Services' discipline system.
21. Office for National Statistics (ONS) Social Trends (1999) website, <www.statistics.gov.uk>.
22. *The Herald*, 30 April 2001: 'Mobile phone chain 4U is to lure high-level graduates into the group by offering them £60,000 per year as a salary. 4U has been hit by a tight recruitment market and is, therefore, launching the Graduate High Flier Scheme.'
23. See <www.henleycentre.co.uk>.
24. Howieson (note 4) p. 5.
25. See the Institute of Employment Studies (IES) Research Club website: <www.employmentstudies.co.uk/research/balance.html> and the DofEE website dedicated to the work-life balance, <www.dfee.gov.uk/work lifebalance/guidance.htm>, 'Work-life balance is for everyone! Whether you are an employer, parent, carer, jobseeker or just someone who wants to achieve a better balance between work and the rest of your life, these sensible policies can benefit you.'
26. See the *Guardian*, 15 September 2001. In the supplement, 'Rise – Next Moves for Graduate Professionals,' it states, 'Nationwide Building Society, which gives staff job-shares, flexible hours, working from home, career breaks, and time off for charity work, has been named the UK's top employer. "Graduates are looking for more than just a high salary and prestige", says a researcher for the Work-Life Balance Trust.'
27. The literature on the Psychological Contract (PC) is extensive. Indeed, the PC may be regarded as the human resources management challenge of this decade. In addition, it is clear that many organisations are now taking this concept very seriously. The Royal Bank of Scotland/NatWest Group undertook a considerable review of the PC with its staff. The programme was designed to end the company's 'job-for-life' culture, which focused on pay for level rather than performance. This was replaced by an achievement culture, where staff were expected to be more proactive within specialist roles.
28. For a full analysis of the PC see P. S. Makin, C. L. Cooper and C. S. Cox, *Organisations and the Psychological Contract* (Leicester: The British Psychological Society 1996).
29. There are many indicators of the 'health' of the PC between the BAFs and its personnel. The following observations – made by the authors – may indicate that a breakdown in the contract has already taken place: high wastage among officers of all ranks, particularly, junior and senior officers; general dissatisfaction with a work environment that is risk averse, punitive and demanding without reward; overt examples of careerism (as opposed to professionalism); career education and training perceived to lack substance and credibility - officers prefer 'credible' civilian qualifications (MBA etc); perceptions that junior officers are not being provided with a value set appropriate to the profession; and complaints that career expectations are not being met. In times of significant retention problems in the BAFs, urgent and detailed research is required involving the PC in a military context.
30. The impact of the IT/IS revolution is now becoming very clear to the MoD. In

the *Future Strategic Context for Defence*, this theme is amplified by two quotes that are very significant: 'Revolutionary changes in technology (or the application of technology) will require matching changes in military doctrine, culture and structures to realize their full potential' and 'Military advantages will rest with those who most identify opportunities, adapt them for military use and integrate them rapidly into equipment platforms weapon systems and force structures.'

31. Edinburgh University cohort: Sociology students; Heriot-Watt University cohort: Management students.

32. The average response rate (for all three Services) for Questionnaire One was 86 per cent. The majority of respondents were in the 18–24 age group (73 per cent). There was a mixture of employs prior to undertaking officer training: 9 per cent of the candidates joined direct from school; 38 per cent of respondents joined direct from tertiary education; 11 per cent of respondents direct from the 'ranks'; and 42 per cent of the respondents came from other employment prior to joining the BAFs. A large percentage of the respondents were male (82 per cent) and virtually the entire cohort was white (UK born and settled).

33. The average response rate (for both universities) for this questionnaire was 26 per cent. The majority of respondents were in the 18–24 age group (92 per cent). There was a mixture of educational backgrounds: 72 per cent of the respondents from Comprehensive/State School and 28 per cent of the respondents from Independent/Private School. For this questionnaire, the majority of the respondents were female (64 per cent) and again, virtually the entire cohort was white, UK born and settled, (92 per cent). Although the sample size and response rate was significantly lower than Questionnaire One, there is a broad representation of respondents, including a mix of locations of birth: Scotland (53 per cent); England (22 per cent); Wales (3 per cent); Northern Ireland (13 per cent); and other (9 per cent).

34. 26 organisations offered interviews including: ASDA stores, BAE Systems, Bank of Scotland/Halifax, BT, Ernst and Young, Reuters, Royal Bank of Scotland/NatWest, Scottish Power, Vodafone, The Foreign and Commonwealth Office and the Scottish Police Force.

35. A total of 17 personnel from the MoD and 3 services were interviewed.

7

Postmodernism to Structure: An Upstream Journey for the Military Recruit?

CHARLES KIRKE
Royal Artillery

It is axiomatic in the British Army that recruits experience 'culture shock' as they make the transition from young civilian to trained soldier, and that they always have done. This chapter is focused at the particular form of this culture shock in the early twenty-first century. The views expressed within it are entirely those of the author and do not reflect official opinion or thought.

As the vast bulk of recruits join the Army in their late teens or early twenties, the civilian milieu from which they make their transition to the Army is contemporary British youth culture. Although this is not a seamless entity throughout the British Isles, there are some constant, or at least very common, features in British youth culture which have at their core certain ingredients first seen in the late twentieth century intellectual movement that has come to be called 'Postmodernism'.

We will first, therefore, describe the postmodernist movement and the resultant elements of British youth culture. Then we will examine the organisational culture of the British Army to assess the gap that recruits have to cross in their transition from one to the other. Finally we will briefly consider the implications for the recruit.

POSTMODERNISM

Like many intellectual movements, Postmodernism arose in reaction against existing trends in thinking. In this case, the ruling paradigm was that of 'modernity', the body of ideas that blossomed in Europe in the eighteenth century but had roots that went back to the Renaissance. Modernity sought to break away from what were seen as the confusions and superstitions of the past and replace them with rationality and objective science.

> The postulates [in modernity] of the thinking self and the mechanistic universe opened the way for the explosion of knowledge under the banner of what Jurgen Habermas called the 'Enlightenment project'. It became the goal of the human intellectual quest to unlock the secrets of the universe in order to master nature for human benefit and create a better world. This quest led to the modernity characteristic of the twentieth century which has sought to bring rational management to life in order to improve human existence through technology.[1]

Important assumptions of modernity and the Enlightenment project were that knowledge could be only be acquired by reasoned rational and dispassionate inquiry, and that this knowledge gave access to fundamental truths that had an independent and lasting existence.

Postmodernism appeared as an identifiable intellectual phenomenon in Western universities in the 1970s, though its roots can be traced back to earlier bodies of thought exemplified by the 'critical theory' movement in the 1930s. Although it is by its very nature diffuse and fragmented into a mass of 'postmodernisms', its general tenets challenge Enlightenment assumptions, questioning the existence of fundamental or objective truth, and suggesting that no aspect of life is fixed or durable.

Instead, human perception is seen to be socially conditioned: what are interpreted as 'truths' are in fact conditioned by the attitudes and expectations prevalent in the observer's society. Creative individuals are therefore free from any assumptions rooted in the past and can express themselves in any way that they feel is right. As Harvey puts it,

> Fiction, fragmentation, collage, and eclecticism, all suffused with a sense of ephemerality and chaos, are, perhaps, the themes that dominate in today's practices of architecture and urban design. And there is, evidently, much in common here with practices and thinking in

many other realms such as art, literature, social theory, psychology, and philosophy.[2]

Sarup amplifies this point,

> Among the central features associated with post-modernism in the arts are: the deletion of the boundary between art and everyday life; the collapse of the hierarchical distinction between elite and popular culture; a stylistic eclecticism and the mixing of codes. There is parody, pastiche, irony and playfulness. Many commentators stress that postmodernists espouse a model which emphasises not depth but surface. … It is also said that in postmodernism there is: a shift of emphasis from content to form or style; a transformation of reality into images; the fragmentation of time into a series of perpetual presents.[3]

However, these freedoms also bring doubts and uncertainties which go far beyond the creative world of the arts and the intellectuals. If what people see as 'truth' is simply a socially conditioned reaction to the way in which they experience the world, then many of the fundamental structures of people's everyday lives are called into question.

Thus Postmodernism has escaped from the world of the intellectuals into popular culture, including that of Britain. We can see its influence in everyday life in various ways. In particular, aspects of our culture that provided frameworks and form to the attitudes and expectations of ordinary people before the 1980s have now been replaced with ephemeral free-formed elements that lack enduring structure.

We can see this effect, for instance, by observing the now established preference for the use of sound bites and pastiche in broadcasting rather than coherent in-depth analysis, the great importance in politics given to 'image' and 'message' which concentrate on the immediate and the superficial, and, arguably, the general reduction in the expectation that marriage will involve a lasting exclusive commitment.

Examples of the major contrasts that exist between the Enlightenment and Postmodernism are given in the following table:

TABLE 1
ILLUSTRATIVE CONTRASTS BETWEEN ENLIGHTENMENT AND
POSTMODERN PARADIGMS

Enlightenment	Postmodernism
Structure	Chaos
Constants	Transience
Objective Truth	Experience
System	Syncretism
Method	Inspiration
Certainty	Uncertainty

In Britain, youth culture can be seen as representing the current extreme of the penetration of Postmodernism in popular culture. External frameworks and order have been replaced by the primacy of individual experience and self-expression. Rules are at best tolerated and questioned, and are often rejected.

Youth music has moved from the limited number of accepted enduring categories of the 1950s and 1960s (such as 'ballad', 'rock and roll', and 'jazz') to a multiplex and fragmented array of changing and in some cases transient categories ('house', 'hard rock', 'soft rock', 'nu-metal', 'heavy metal', 'dance', 'hip-hop', 'techno', 'rap', 'reggae', 'garage', and so on). When it is performed or broadcast it is frequently combined with visual images that have no particular connection to the music and are presented in an apparently random and unconnected stream. Dancing has no structure at all – or rather, it has the structure of the instant, the context, and the inspiration of the moment.

Drugs, the ultimate celebration of experience and illusion over the mundane facts of everyday life, are perceived by many young people as a legitimate form of recreation.

Religion, for those who practise it, often follows the New Age patterns of self-realisation through experience rather than worship of a constant, powerful, and loving external entity.

Sex is a legitimate form of recreation rather than part of a long-term bonding process. In short, life is lived in the experience of the moment, with the transient structure of the moment and any idea of permanence or overall structure is irrelevant.

These aspects have been brought out clearly by Cray, in *Postmodern Culture and Youth Discipleship*, in which he focuses on the social and cultural forms which shape young people's lives and expectations. For him, 'One of the many paradoxes of the postmodern world is that there is a great emphasis on image, appearance and style. "Enjoy the surface' is a piece of postmodern wisdom. ... The postmodern self tries to construct its own continuously changing centre.'[4]

This then is the environment which the majority of recruits consider natural and normal when they join the Armed Services. As the research which underpins this chapter was carried out almost exclusively in the British Army, it is to that institution and its organisational culture that we will turn now.

SOCIAL STRUCTURE

The analysis which follows is based on the concept of 'social structure'. The idea of 'social structure' is in essence a framework for everyday life to which all integrated members of the society or human group in question subscribe. It is expressed in the regularities of the day to day activity of those people where,

> the events which comprise human behaviour exhibit regularities whose forms are mutually interdependent, over and above their interdependence in the personality-behaviour systems of each individual actor.[5]

Giddens has put it more succinctly as 'some kind of patterning of social behaviour', adding that,

> As ordinarily used in the social sciences, 'structure' tends to be employed with the more enduring aspects of social systems in mind ... . The most important aspects of structure are rules and resources recursively involved in institutions.[6]

For the purposes of this chapter, therefore 'social structure' will be used in the following sense: a body of ideas, rules and conventions of behaviour which governs how groups of people or individuals organise and conduct themselves vis-à-vis each other. Conceptually it therefore provides the indispensable background to, and framework for, daily life.

It is now widely accepted that the concept of 'social structure' provides a static image of a human group, which includes an implicit assumption that individuals automatically subscribe to it. The concept therefore needs to be balanced by a consideration of the dynamics of everyday life: individuals make their own decisions about how to behave. Indeed, in their behaviour, or the processes of everyday life, they reproduce or develop the underlying assumptions of their lives, and thus by their practice they have an effect on the social structure. This has been captured by Giddens in the term 'structuration', which is,

> best thought of as a useful term designating the process of expression and reproduction of social structure (or structural systems) in the informed behaviour of agents who draw upon rules and resources in the diversity of action contexts.[7]

A useful analogy by which social structure may be distinguished from process is that of a map. Social structure provides a cognitive map of the social terrain in which the individual finds himself or herself, and that individual navigates his or her own path through it in the process of living. Such a concept proposes the existence of a common structured body of rules and shared expectations within a cultural group while still allowing scope for individuals to act in ways of their own choice, and sometimes even to amend the map. In this sense, the model set out below describes the common map which is available to soldiers: each will find his or her own way through the terrain represented by it.

THE ORGANISATIONAL CULTURE OF THE BRITISH ARMY

This section briefly examines the new rule-set, the new social map, which the recruit confronts in his or her first few weeks as a soldier. In particular, it presents a summary of a model which seeks to capture the organisational culture of the British Army. This model was first published in 2000[8] and has since been developed slightly. It has been constructed during one year's full-time research under the MoD Defence Fellowship Scheme followed by five years research as a part-time PhD candidate with Cranfield University (RMCS). The principal subject for the research was the observed behaviour of soldiers of all ranks from private soldier to lieutenant colonel at regimental duty, and their experiences and attitudes expressed in 119 individual interviews and a small number of focus group sessions.

The main field within which the research was conducted was units of the 'combat arms', that is to say those which can be expected to be involved as formed units in the forward battle area in conventional operations. These units comprise the Household Cavalry, the Royal Armoured Corps, the Royal Artillery, the Infantry, the Royal Engineers, the Royal Signals, and the Army Air Corps. However, strong indications emerged during the research that the attitudes, expectations, and behaviour encountered in those arms are generically similar to those of remainder of the British Army. Certainly for the purposes of this chapter we may consider them representative of the overall organisational culture which the recruit enters when he or she joins any part of it.

Initially, the work was aimed at identifying the 'social structure' of life at regimental duty in the Army. However, it soon became clear that no single framework could be developed to capture a body of ideas, rules and conventions of behaviour which matched the observed behaviour of soldiers. Different contexts seemed to demand behaviour that was substantially different from that in other contexts. Attention was therefore given to the identification of the range of possible contexts at regimental duty.

Four different families of contexts were identified, each with appropriate behavioural frameworks. These were identified as four separate but contiguous social structures, which were brought together to form the top level of a model that provides a powerful means to describe, analyse and predict soldiers' behaviour. This top level is sufficient for the purposes of this chapter, representing as it does a generic insight into the organisational culture of the Army. However, it should be born in mind that investigation of more specific areas of that culture would involve using greater degrees of complexity in the model.

As far as the observed behaviour of the soldiers was concerned, therefore, this model seeks to capture the fact that soldiers' behaviour differs in different contexts, but these contexts can be described satisfactorily by a minimum of four categories, within each of which the expected patterns of behaviour are broadly similar. We may call these four categories of contexts, four 'social structures'.

The Four Social Structures

It was found analytically convenient in constructing the model to ascribe specific meanings to certain terms. Whenever these words are used in this chapter, therefore, they are printed in *italics* and defined on first use.

The four social structures comprise:

1. The *formal command structure*, which is the structure through which a soldier at the bottom receives orders from the person at the top. It is embedded in and expressed by the hierarchy of rank and the formal arrangement of the unit into layer upon layer of organisational elements. It contains the mechanisms for the enforcement of discipline, for the downward issue of orders and instructions and for the upward issue of reports, and it provides the framework for official responsibility.

2. The *informal structure*, which consists in unwritten conventions of behaviour in the absence of formal constraints, including behaviour off duty and in relaxed duty contexts. An important element in this structure is the web of informal relationships within the unit. Individuals come into personal contact with other people within the unit, of any rank, and establish inter-personal relationships with them.

   Although it might appear at first sight that the quality and intensity of such relationships are determined by free choice on the part of the individual (because they are informal), the network of a soldier's informal relationships is for the most part constrained by his rank and position in the unit.

3. The *loyalty/identity structure*, which is manifested most obviously in a nesting series of different sized groups which are defined by opposition to and contrast with other groups of equal status in the *formal command structure*. The structure itself, the 'body of ideas, rules and conventions of

behaviour', consists in the attitudes, feelings and expectations of soldiers towards these groups and their membership. Thus an infantry soldier would express his identity as a member of his platoon and feel loyalty to it in competition with other platoons of the same company.

However, where his company is in competition with other companies, these attitudes and feelings would be transferred to the company, rather than the platoon, and this process is continued up to levels beyond the unit (and down to those below the platoon).

4. The *functional structure*, which consists in attitudes, feelings and expectations connected with the carrying out of specific tasks and military activities. Where groups are formed to carry out such functions, they might exactly reflect the *formal command structure* (which provides an easy and quick means of creating any group within a unit) or they might be independent of it. For example, an infantry platoon (a basic element in the infantry command structure) tends to carry out military functions on exercise and operations as a formed body.

In contrast, a 'rear party' which remains in barracks while the rest of the unit is away (perhaps on leave or on an operational tour of duty) is usually made up of soldiers from all over the unit, brought together into an *ad hoc* grouping.

These four social structures are illustrated in Figure 1.

FIGURE 1
THE FOUR SOCIAL STRUCTURES

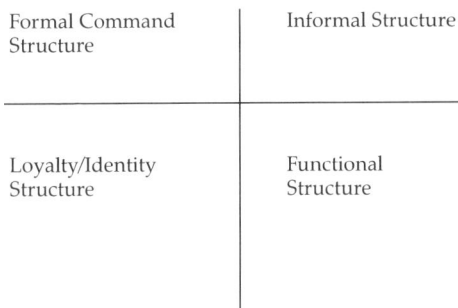

Formal Command Structure	Informal Structure
Loyalty/Identity Structure	Functional Structure

Other Elements to the Model

The four *social structures* provide the core of the model, but to make the model practicable in use certain qualifying elements need to be added.

First, it is explicit in the model that soldiers only exercise one social structure at a time, though the transition between different social structures may occur rapidly as the context changes. Take, for example, this typical vignette:

> *Observation*: An officer approaches a group of soldiers who are relaxing near their vehicle during a break from maintaining it. The senior member of the group brings it to attention and salutes the officer. The officer returns the salute, and tells them to 'carry on'. The group relaxes and a few minutes later the members return to their vehicle maintenance.
>
> *Analysis*: The soldiers are exercising the *informal structure* as they relax. The approach of the officer necessitates a change to the *formal command structure*. He or she returns them to the *informal structure* by telling them to 'carry on', and they subsequently move to the *functional structure* when they return to work.

The structure of the moment is named in the model as the *'operating structure'*.

It may be argued that it is unrealistic to insist upon the modelling of one single *operating structure* at any one instant, and it must be accepted as a possible artificiality in the model. However, it remains a useful device in practice.

First, it matches a very high proportion of the soldiers' behaviour observed during the research: as we have just seen, their behaviour does indeed change with the context of the moment.

Second, this assumption encourages the observer to look for the moments of transition between structures, even in ambiguous and confusing situations, and thus detect subtle changes in context.

The second qualifying element is a constant factor which exists in all the social structures, which is best called *'superiority and inferiority'*.

Each social structure has an embedded idea of hierarchy: for the *formal command structure* it is rank; for the *informal structure* there are informal hierarchies of power and prestige (see Killworth[9] for example); for the *functional structure* there is the importance given to the variability between individuals in their military skills and their ability to carry out military tasks; and in the *loyalty/identity structure* each element assumes itself 'the best' in some way.

Third, it should be borne in mind that although the four *social structures* in the model have been described separately, they do not exist in isolation from each other. They inform each other in matters of detail and are intertwined and overlaid in an intricate and complex pattern. The resulting interconnections, which are experienced in the soldiers' daily lives, engender a flexibility and suppleness in the social system that under normal circumstances prevents insurmountable structural barriers arising between individuals or groups within a unit.

Informal Relationships

The *informal structure* seemed to be the most chaotic and complex of the four because its rules and conventions of behaviour are generated and reproduced entirely by the mutual consent and cooperation of those operating within it. However, the patterns of informal relationships that emerged during the research were so clear that it was possible to construct the following typology, which represents a sub-model in its own right.

Five identifiably different types of informal relationship were identified, as follows:

1. *Close Friendship.* As defined in the model, *'close friendship'* consists in a durable relationship that transcends the military environment, where there is a large measure of trust and respect between the parties and few barriers to discussion of highly personal matters. In interviews with soldiers of all ranks it was established that, for virtually every one, a useful test to identify *close friendship* would be to determine whether the relationship would survive unchanged if one of the parties was prepared to shed tears in the presence of the other. It is a rare and special relationship.

149

In the words of a warrant officer in an Infantry battalion 'I've maybe made only two or three close friends in my career, though I've had plenty of military friends.' This rarity is an important feature. It is sufficient to recognise existence of the relationship, but we must also acknowledge that it is sufficiently scarce that it is not a regular feature of regimental life for many individuals.

2. *Friendship.* The term *'friendship'* is used specifically in the model to refer to a less intense relationship which is frequently found to exist between soldiers within the *informal structure.* It can have all the appearance of *close friendship,* in that individuals constantly seek each other's company, will help each other if they are in trouble, and will be prepared to share almost anything if the need arises, but it falls short of the depth and intensity of the other relationship.

Thus, during an interview one soldier said of his particular circle of mates that he would be more than prepared to help any one of them: if a bloke was feeling unhappy then his friends would naturally take him out drinking to cheer him up. However, if a mate wanted to discuss deeply personal matters then he 'would not want to know!'

Bonds of *friendship* are usually formed within narrow bands of rank. For example, private soldiers may form *friendships* with lance corporals, but a *friendship* with a full corporal may attract disapproval. Similarly, warrant officers may form *friendships* with sergeants that they have known for some time, but there will always be a certain distance in the relationship (particularly if they are in the same sub-unit).

Senior lieutenants may form *friendships* with captains, but junior second lieutenants are unlikely to do so. Although there are no formally stated regulations which proscribe *friendships* growing up between people of widely diverse rank, such relationships are frowned upon because they are held to be potentially compromising for discipline.

3. *Association.* It is often found that two soldiers separated by rank distance wide enough to exclude *friendship* between them will come into regular contact and will form an informal bond of mutual trust and respect that falls short

of *friendship* as defined above, but is nevertheless an important bonding feature. Such a relationship will probably arise, for example, between an Infantry platoon sergeant and his platoon commander, and adjutant and his or her chief clerk or between an Artillery battery sergeant major and his or her battery commander. This relationship was given the name '*association*' in the study.

4. *Informal Access.* It is recognised, though not officially laid down, that each individual has a right to speak informally and without a formal appointment with certain other people who are at a certain degree of structural distance (superiors in his chain of command for instance), even though a link of *association* does not exist between them. Thus a junior officer can expect to be able to have '*informal access*' to his sub-unit commander, as a private soldier can to his platoon or troop commander. Similarly, any member of a Sergeants' Mess can expect to have opportunities informally to approach the Regimental Sergeant Major.

5. *Nodding Acquaintance.* The term '*nodding acquaintance*' encompasses all the informal relationships which are not encompassed by the other terms. In essence, it is a relationship where the parties know each other by sight, but not necessarily by name, and they acknowledge each other's existence and common participation in the same segment of the *formal command structure*. The relationship may remain as it is, or it may grow into any one of the others listed above.

A diagram illustrating these relationships is at Figure 2. 'Ego' represents an individual who is somewhere in the middle of the rank structure (somewhere between sergeant and captain) and who therefore has informal relationships with those senior and junior to himself or herself.

In reading this diagram it is important to note that the spaces between the boxes are voids. They show either areas where potential relationships do not exist (as, for example, a rank-based relationship – *association* or *informal access* – between peers) or they are there simply to differentiate between the boxes (as in the gaps between *informal access* and *association*).

In the horizontal axes of the boxes, the diagram allows for different degrees of closeness within each box: the further to

FIGURE 2
INFORMAL RELATIONSHIPS

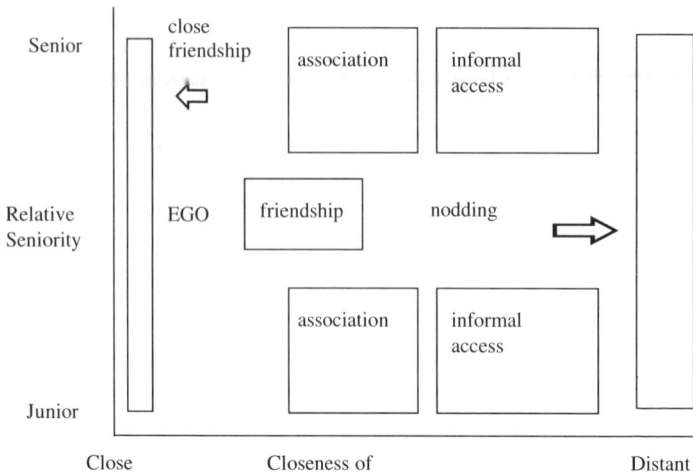

the left the stronger. This reflects the fact that, apart from *close friendship*, which is by definition a strong mutual bond, and *nodding acquaintance*, which is essentially weak, a significant variable in any particular case is the strength of the relationship.

The Nature of the Model

During the research it became clear that the organisational culture of the British Army is particularly amenable to analysis by the identification of rule sets, and that these rule sets fell into only four separate definable categories. There was some concern that this was an artefact of the research, but it was eventually concluded that it actually reflected the fact that the existence of such rule sets was indeed accepted as part of the organisational culture by its participants.

It was also observed that there is an inherent tendency for soldiers to want to know what is 'appropriate' and what is 'right' and to expect such behaviour from their fellow-soldiers. Individually and corporately they find themselves articulating this tendency in their daily lives. Examples are easy to find.

The *formal command structure* encompasses unambiguous rank structures, publicly displayed in such things as badges of rank, and written codes of behaviour – *Queen's Regulations*,[10] the *Manual of Military Law*,[11] the *Drill Manual*,[12] and unit-authored Daily Routine Orders.

The *informal structure* includes a significant level of pressure to 'fit in' with one's peers, subordinates and superiors, in an appropriate part of a rank-dependent array of informal relationships as captured in Figure 2.

Functional prowess is an important feature in the *functional structure* and soldiers openly judge each other's value by their functional attitudes and standards of achievement, and units lay down the way that many operational and peacetime tasks should be carried out in Standing Operating Procedures.

The *loyalty/identity structure* appears clearly delineated in the organisational structure of the unit and, above the unit, in distinctions of dress and in unique unit customs and artefacts, and the soldiers articulate loyalty/identity values in their actions and conversation.

Although the model was derived entirely from participant observation in the British Army and from interviews with soldiers, as the investigation evolved, it was found to have something in common with other concepts in social science. In the first place, the rules sets identified had something in common with the use by Giddens' of Goffman's concept of 'frames'[13]:

> Frames are clusters of rules which help to constitute and regulate activities, defining them as activities of a certain sort and as subject to a given range of sanctions. Whenever individuals come together in a specific context they confront (but, in the vast majority of circumstances, answer without any difficulty whatsoever) the question, 'What is going on here?
>
> What is going on?' is unlikely to admit of a simple answer because in all social situations there may be many things 'going on' simultaneously. But participants in interaction address this question characteristically on the level of practice, gearing their conduct to that of others. ... Framing as constitutive of, and constricted by, encounters 'makes sense' of the activities in which participants engage, both for themselves and for others.'[14]

Second, the idea of a limited number of such rule sets that interacted in a complex system had common ground with Geertz's concept of 'planes of social organization' in Bali. These 'planes' consisted in a set of seven interacting and intersecting organisational components which between them could be used to account for form and variation in Balinese village social life.[15]

Although the model's construction and application to the British Army are novel, therefore, it remains within the broad main stream of social scientific thought.

IMPLICATIONS FOR THE RECRUIT

It will by now be fully apparent that the organisational culture of the British Army is entirely alien to the youth culture described earlier. Where this youth culture lacks any idea of solid structure, contains an assumption that all aspects are ephemeral, and values feelings and experience over external standards, British Army organisational culture is highly structured, based on clear shared rules (both written and unwritten), and contains a constant assumption of hierarchy. In essence, where youth culture is postmodern, British Army organisational structure is anything but.

This, then, is a key element of the 'culture shock' experienced by recruits to the British Army. They have to make a considerable cultural leap, greater than their forebears who came from the more structured youth culture of the past[16] and this leap concerns their basic expectations of life and their deeply held assumptions and attitudes. To make matters more difficult for them, they have to do this in an environment, the Army Training Regiment, which is controlled by staff who tend to be highly socialised into the culture that the recruits are trying to join.

Currently, no allowance appears to be made for the difficulties which many recruits must experience in making this cultural transition, and it is entirely possible that a proportion of the numbers currently lost during recruit training are lost primarily because of such transition.

It would seem sensible therefore to consider incorporating this factor in the structure of the recruit training programme. Instead of demanding that individuals leap directly from their familiar cultural milieu into the alien culture of the Army, there may be considerable benefit in deliberately creating a bridge between the two and helping them to cross it in the early weeks of their training.

No quick answer can be offered here as to how such a bridge might be constructed. Its design would be a matter for future work, but however it is done, it will be important for the designer to have a good appreciation of both contemporary youth culture and Army organisational culture, the two ends of the bridge. It is hoped that this chapter has contributed to the process.

REFERENCES

1. Grenz S. *A Primer on Postmodernism* (Cambridge, UK: William B. Erdmans 1996) p.16.
2. Harvey D. *The Condition of Postmodernity: An Enquiry into the Origins of Cultural Change* (Oxford: Basil Blackwell 1980) p. 98.
3. Sarup M. *An Introductory Guide to Post-structuralism and Postmodernism,* (2nd ed.) (London: Harvester Wheatsheaf 1993) p.132.
4. Cray, G. *Postmodern Culture and Youth Discipleship: Commitment or Looking Cool?* (Cambridge: Grove Book 1998) pp.17 and 18.
5. Gluckman, M., and Eggan, F., 'Introduction', in Banton M. (ed.) *The Relevance of Models for Social Anthropology* (London: Tavistock Publications 1965) p.xviii.
6. Giddens, A. *The Constitution of Society: Outline of the Theory of Structuration* (Cambridge: Polity Press 1984) pp.16 and 23–4.
7. Ibid. p. 16.
8. Kirke C. 'A Model for the Analysis of Fighting Spirit in the British Army', in Strachan, H. (ed.) *The British Army, Manpower and Society into the Twenty-first Century* (London and Portland, OR: Frank Cass 2000) pp.227–41.
9. Killworth, P. *Culture and Power in the British Army: Hierarchies, Boundaries and Construction,* Cambridge University PhD Thesis, 1997.
10. MOD. *The Queen's Regulations for The Army 1975 (Including Amendment 24)* (London: TSO [The Stationery Office] March 2001).
11. MOD. *Manual of Military Law Part I, 12th Edition* (London: TSO, 1 May 2001).
12. MOD. *The Drill Manual*. D/DAT/13/28/97. 1990.
13. Goffman, E. *Frame Analysis: An Essay on the Organization of Experience* (Harmondsworth: Penguin Books 1974).
14. Giddens (note 6) p.87.
15. Geertz, C. 'Form and Variation in Balinese Village Structure', *American Anthropologist* Vol.61 (1959) pp.991–1012.
16. Cray (note 4) pp.3–4.

8

AFOPS, Work-Life Balance and the Problems of Recruitment and Retention

ANDY BOLT
Royal Air Force

The concept of work-life balance has become prominent in recent years, with the government launching a £5.5 million Work-life Balance campaign in March 2000.[1] Indeed, work-life balance is becoming employers' top priority,[2] being a consequence of strategic human resource management[3] (HRM) where people are central to an organisation with competitive advantage gained by maximal concentration on HRM issues.

At the same time, the armed forces have been subject to a series of fundamental reviews and are implementing HRM initiatives, headed by the Armed Forces Overarching Personnel Strategy (AFOPS).[4] The success of an armed forces' HRM strategy and work-life balance is key to the organisation's future, particularly against a backdrop of recruitment and retention problems.

This chapter will explore the applicability of work-life balance principles to the armed forces. It will then consider whether AFOPS will make a difference to armed forces' work-life balance and whether recruitment and retention can be improved by promoting work-life balance policies.

WORK-LIFE BALANCE

There is a growing body of literature on work-life balance from a variety of perspectives: theoretical,[5] empirical[6] and practical, with organisations such as the Department for Employment and Equal Opportunities,[7] Trades Union Congress[8] and Department for Trade and Industry[9] providing comprehensive information. A full literature review is beyond the scope of this chapter, but the headline concerns are:[10]

- The organisation of working life over the life course
- Time for caring and families

- Time for learning
- Time for ourselves
- Time for community/political participation
- Social, personal, economic and political time: a new politics of time

A primary focus has been time based: there is evidence of a growing dissatisfaction in the number of working hours[11] with 40 per cent of people wishing to work a different number of hours.[12]

The effect of work-life imbalance is demonstrated in the cost to British businesses of £10 billion per annum due to work-related stress.[13] Furthermore, with an increased intensity of work,[14] 'burnout is reaching epidemic proportions'.[15] Some 42 per cent of all workers feel 'used up' by the end of the day and 69 per cent would like to live a more relaxed life.[16] However, 74 per cent of employees believe they have the right balance.[17]

The business and economic case for a work-life balance is generally established, with evidence that work-life balance policies increase retention and bring benefits such as happier staff.[18] However, some studies do question the veracity of this conclusion,[19] possibly because many organisations pay lip service to such policies.[20]

More cynically, Craig and Kimberley suggest that, 'the work-life balance campaign merely allowed employees to admit that they had lives outside work. Admitting that personal lives might sometimes take priority over their professional lives, however, was still taboo.'[21]

HUMAN RESOUCE MANAGEMENT AND THE ARMED FORCES CONTEXT

As part of the myriad of reviews and studies that have been brought to bear on the British armed forces in the last 15 years, personnel issues have gained prominence. This has mirrored the general move from financial to technological to human capital values.[22] Indeed, the World Bank's Wealth Index defines the wealth of nations as primarily (60 per cent) 'human capital'.[23]

Employers now pay more attention to the well-being of their workers[24] and there has been a change in emphasis from people being made productive by the system, to people enabling a productive system.[25] Furthermore, people are less willing to display unlimited commitment to their employer, partially in response to the reduction in employers providing secure progressive careers.[26]

Implementation of Work-Life Balance Policies in the Armed Forces

As well as reflecting the growing recognition of HRM, the MoD's policies have contained work-life balance elements. Both the Policy for People[27] and the Defence Mission[28] explicitly recognise the importance of families to the success of the armed forces. The logical inference is that Service personnel do not have a clear delineation between work and personal life and that the organisation is responsible for helping the individual achieve a sustainable work-life balance.

The next stage from the Policy for People was the Armed Forces Overarching Personnel Strategy (AFOPS), which aims to provide direction backed by a comprehensive action plan. Thus the armed forces' 'organisational and business needs can be translated into coherent and practical policies and programmes'.[29]

This means that even if AFOPS is a positive initiative it must be accompanied by more detailed practices if it is to reap the benefits desired. However, a comparison of the work-life balance policies available within the Services with the those practised elsewhere shows limited implementation within the armed forces.

Recruitment and Retention

The armed forces are currently failing to achieve their annual recruitment target of 25,000[30] people and 'armed forces manning levels remain fragile'.[31] The question is whether this problem is due to a work-life imbalance or to other factors. For instance, there is a declining recruitment pool caused by demographic changes; or the perceived lack of a clear threat to the UK may lessen the attractiveness of the armed forces by reducing the sense of public service. Furthermore, expeditionary operations have little apparent relevance to protecting the UK, but demand greater privations.

Retention is also poor, with increased outflow of personnel from the RAF and Army.[32] 'Many do leave because they are

TABLE 1
WORK-LIFE BALANCE POLICIES

Type of Policy	Examples	Available to armed forces personnel
Flexibility in total time worked	• Reduced working hours • Part-time working • Job sharing	No
Flexibility in when work occurs	• Annualisation of hours • Flexi-time • Self-scheduling and team 'self-scheduling'	No
Breaks in work	• Career breaks and sabbaticals • Parental leave	Yes (e.g. MDA)
Flexibility in type of work	• Job rotation	Postings
Flexibility in work location	• Homeworking, tele-working	No
Situational support	• Childcare • Eldercare • Maternity benefits • Dental and Health care • Life-support e.g. laundry facilities, catering • Physical Fitness e.g. gym • Stress management and employee well-being	No No Statutory only Yes Some Yes No

Source: Adapted from Papalexandris N. and Kramar R., 'Flexible Working Patterns: Towards Reconciliation of family and Work', *Employee Relations*, Vol. 19, No. 6 (MCB University Press 1997), pp. 581–95; and Pillinger J., 'Work/life Balance: Finding New Ways to Work', <www.tuc.org.uk/work_life/tuc-4022-f0.cfm> accessed 31 January 2002.

not satisfied with the quality of life the Services offer them',[33] whilst others see continual change and redundancies as a betrayal by the armed forces. Why dedicate oneself to the Services in an environment of manpower reductions, over-stretch and low morale?[34]

The situation is such that the Armed Forces' Pay Review Body made an unprecedented move by including a chapter on Quality of Life[35] in their 2001 Report. Similarly, the Services' Continuous Attitude Survey has pointed to work-life balance issues such as the effect of separation on family life, leave restrictions and hours worked as influencing retention.

Recruiting and retaining suitable personnel necessitate matching organisational needs and priorities with those of the target population.[36] One fundamental issue facing the military is the perception that 'many recruits today did not instinctively share the core values of the Service.'[37] The armed forces recruit almost exclusively from a young age group. Providing a suitable work-life balance for this demographic is difficult because at that stage of life many people are still resolving parental, peer group and personal goals.[38] Their needs are therefore undefined and likely to change.

Social change has forced the military to widen its target demographic to include women and to offer continued service to pregnant women whose needs were previously seen as incompatible with those of the service. Furthermore, the armed forces must reflect the diversity of modern society. These changes alter the potential work-life balance elements that the forces must consider. It is noteworthy that the work-life balance concept has largely developed in affluent European and North American countries,[39] against a backdrop of 'selfish individualism'.

How this translates into other societies and cultures, such as the Muslim regard for 'Allah's will' as the force behind all actions and the Buddhist elevation of mental or spiritual development above that of material development is increasingly important in a globalised society and culturally diverse recruiting pool.

Reserve Forces

An interesting juxtaposition comes from the Reserve Forces, whose existence is predicated on the willingness of people to be a reservist in addition to their prime employment. Reserve personnel are therefore either seeking to extend their working hours because their primary employment does not offer them sufficient work or because they regard the Reserves as a worthwhile 'leisure' activity that provides balance with their primary work.

The poor attendance resulting from the compulsory call-up of reservists for Afghanistan may be taken to indicate that the personnel see the Reserves as a leisure activity with a commitment that does not extend to actual operations. This has profound implications for the armed forces, as organisational needs are not aligned with personal needs.

Public Service and Military Uniqueness

All military personnel are expected to display high levels of selfless commitment, courage and discipline. This necessitates a psychological contract that goes beyond the normal bounds of simply earning a living. To strengthen the individual's ability to provide this commitment the armed forces have traditionally sought to create their own inclusive society and to provide a life rather than just a job. The military lifestyle also facilitates and rewards the individual for the extra-ordinary demands, that the military must necessarily make, by providing benefits such as Mess life and adventurous training. A key concept in establishing the bounds of the armed forces span of influence over life as well as work is that the armed forces are 'not *just* here to provide attractive employment opportunities'[40] but to defend the nation.

The military lifestyle is embodied in Tonnies' *Gemeinschaft*,[41] where membership is self-fulfilling and the military is more than the workplace but is a form of family. In this context, work-life balance is internal to the military. Membership cannot be sought simply on the basis of earning a living, but it must be based on the shared beliefs and social unity that are requisite for a family. This is much harder in today's individualistic society where family values have lessened and personal sovereignty[42] is sought.

Institution or Occupation?

The Bett Report[43] documented the move of the armed forces from an institution to an occupation. Again this follows societal trends away from relationship-based job-for-life to transaction-based employment[44] and portfolio careers. Whether work is our defining aspect, the place where people find fulfilment and meaning[45] is fundamental to how work-life balance is viewed. This is especially true in the public sector where sense of service may be an individual's defining motivator.

FIGURE 1
THE PSYCHOLOGICAL CONTRACT

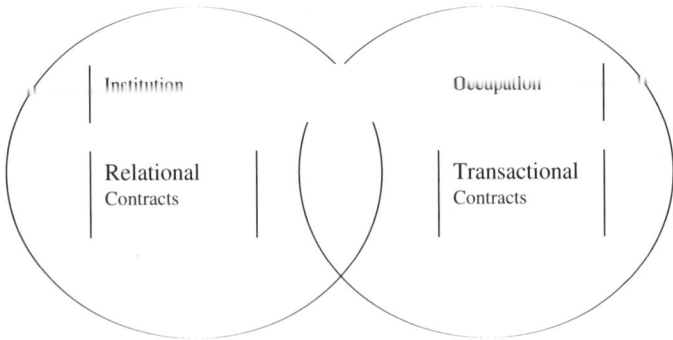

Source: Bristow C., 'The Psychological Contract', in Alexandrou A. and Bartle R. (eds.) *Human Resource Management in the Public Sector: A Uniform Perspective?* (Swindon: Cranfield University, 1999), p. 7.

The relationship model (Figure 1), emphasises feelings of belonging and an identity that is tied to the workplace (institution) whereas the transaction model reduces work to a transient element of life (occupation).

As an occupation the military are concerned largely with the work half of the equation. As an institution, the military's influence covers many aspects on wider life and hence both sides of the equation. Thus, the military must be conscious of their culture[46] and the psychological contract[47] when addressing work-life balance.

The increasing use of agencies and contractors, particularly closer to the battlefield, undermines the traditional perception that military effectiveness requires total immersion in the military lifestyle. If a contractor does not need to be institutionalised why does a soldier?

FROM TWO TO FOUR DIMENSIONS

A Two-Dimensional Model

The concept of work-life balance implicitly assumes a separation of work and life and that work is a negative phenomenon. Work is merely an economic necessity enabling people to gain access to the more important, positive (non-work) aspects of life. Long working hours impinge on home life while personal activities, such as child-care, impinge on work. This is a very two-dimensional approach.

By contrast, the institutional model, avoids this negative connotation of work. Rather than separating and then balance these activities, this model stresses a broader aim – one of 'life satisfaction'. Occupationally, this may be relatively simple to achieve for knowledge workers who contribute mental effort to their employer, but it is much more complex for physical and team tasks which form the vast majority of military activity.

Guest's Three Dimensions

Guest takes the issue into three dimensions by considering individual, life outside work and work:

FIGURE 2
WORK-LIFE INFLUENCES

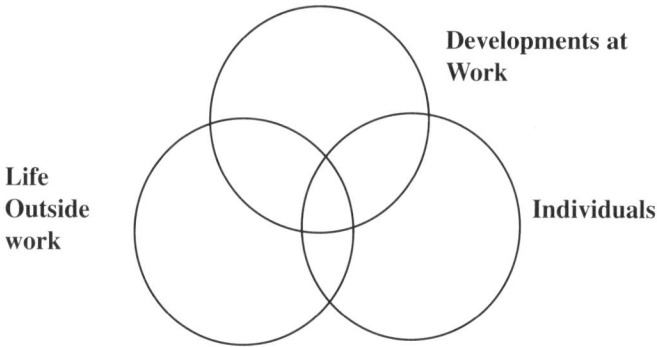

Source: Adapted from Guest D., 'Perspectives on the Study of Work-Life Balance', A Discussion Paper prepared for the 2001 ENOP Symposium, Paris, 29–31 March 2001.

Handy's Types of Work

Charles Handy's personal move from paid employment to work as an independent consultant caused him to analyse how he wished to spend his time.[48] He categorised his 'work' activities into four areas: paid work, study work, home work and gift work. He then planned a balanced allocation of these areas. His view was that paid work provides the means to life and that study work is essential for development of skills, and therefore both longer-term employability and enjoyment. Home work provided the fabric of living and relationships, while gift work provides an altruistic sense of meaning to life.

Here, this concept has been adapted into a framework that can represent, graphically, a range of work-life balance scenarios. According to the range of motives that might drive individual work-life behaviour, different models can be produced, as illustrated here. (Note that a fourth possible model, representing an individual who need not work is not included.)

The person who works primarily to earn a wage has the plot below:

FIGURE 3
WAGE SLAVE (THE 9 TO 5 WORKER)

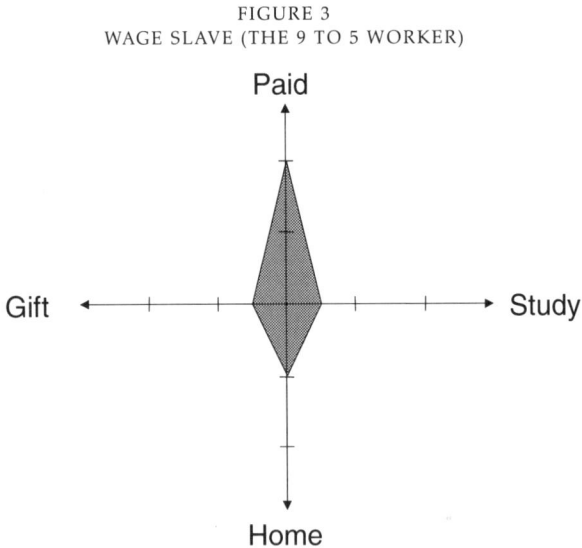

Source: Author.

An individual with a public service ethos has the following plot:

FIGURE 4
PUBLIC SERVICE ETHOS

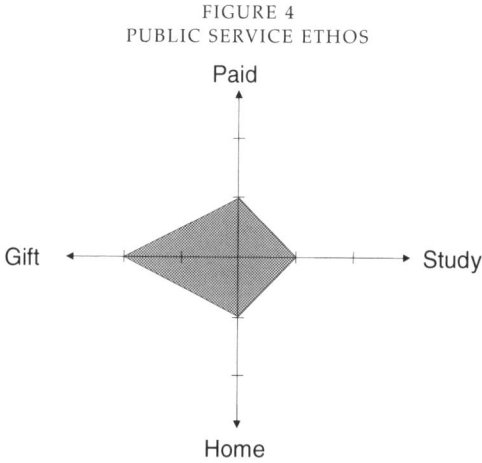

Source: Author.

One of the motivation for joining the RAF is the opportunity to 'learn a trade'.[49] This plot would be represented thus:

FIGURE 5
LEARN A TRADE

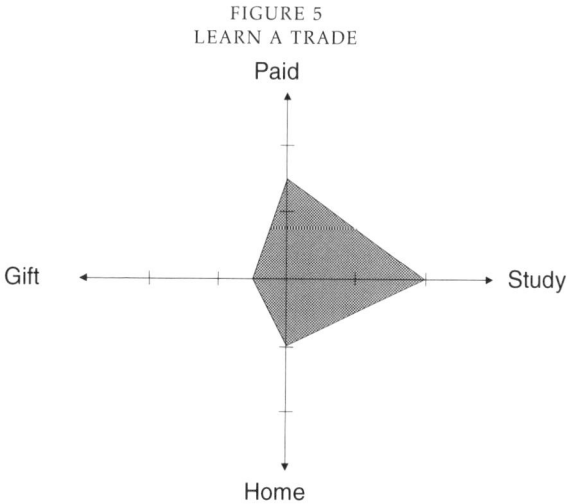

Source: Author.

Taking this one stage further, the work-life offered by the armed forces can be compared to the expectations of its members. Any discrepancies can then be resolved and both HRM policies and recruitment can be tailored to the target population.

FIGURE 6
COMPARISON OF TARGET POPULATION EXPECTATION WITH ARMED
FORCES WORK-LIFE BOUNDARY

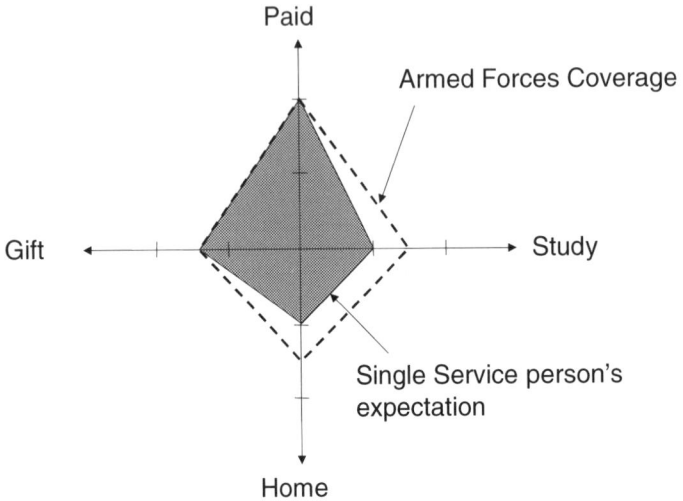

Source: Author.

However, the balance that attracts young single people is not such a powerful motivator for retaining married personnel. Furthermore, while Service personnel themselves gain satisfaction and comradeship from their service, spouses and children may not experience the same level of inclusiveness and hence satisfaction, particularly with the rise in dual career households,[50] which can further separate the spouse from the service lifestyle. The outflow of married personnel is often due to problems experienced by their families.[51]

FIGURE 7
CORRELATION BETWEEN BALANCE OFFERED BY
ARMED FORCES AND EXPECTATIONS OF MARRIED AND SINGLE
PERSONNEL

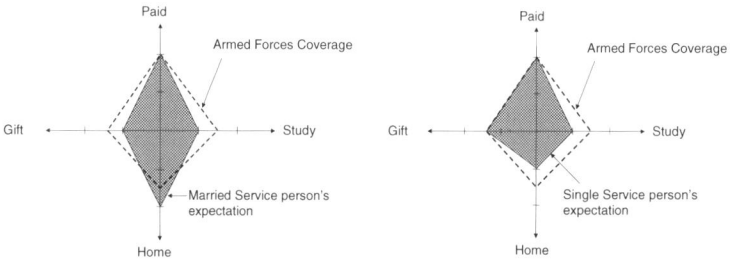

Source: Author.

This analysis is based on armed forces personnel having a basic need for money, but also having a public service ethos that reduces as the demands which a family gives rise. Families also increase the need for home work, conflicting with the provisions of the Services. The opportunity for training,[52] particularly in the RAF, with regular study throughout a Service career, is generally greater than the expectations of members.

The main difficulty with the above analysis is in quantifying each parameter. The 'soft', subjective nature of the importance of each parameter means that the development of a quantitative methodology is difficult. The extrinsic and intrinsic aspects of motivation, complicate this further and must be considered. Nevertheless, the model presented here provides the basis of a new research agenda to challenge those who are concerned about the recruitment of people who will give worthwhile and, potentially, career-long service and those who recognise the need to retain the expertise and knowledge of trained personnel.

TABLE 2
MOTIVATORS FOR WORKING

Reward	Extrinsic	Intrinsic
Description	Tangible rewards, hygiene factors	Opportunity to satisfy other goals
Motivator	Necessity	Job satisfaction
Examples	Paid work, working conditions	Lifestyle (home work), sense of achievement, companionship, status, public service (gift work), challenge & development (study)

Adapted from text in Hunt J. W. *Managing People at Work*, 3rd edn. (Maidenhead, UK: McGraw-Hill 1992), p. 11.

CONCLUSION

The nature of work, the recruitment pool and the nature of the armed forces have all changed and are likely to continue changing. Military HRM policies must be cognisant of this change.

While the concept of work-life balance can be applied to a standard '9 to 5' occupation, it is not an optimal perspective for organisations where membership is institutional and which offer benefits on both sides of the inherently two-dimensional work-life balance equation. Hence, a broader concept of life satisfaction is more useful. The extent to which work or home is a central life interest is then considered along with other factors such as the need for learning and gift work. This four-dimensional life interest has been analysed graphically, allowing comparison of the rewards offered by an organisation against the needs of the individual at various stages of life.

Retention can be improved by fitting the rewards offered to the needs of those that are leaving. Recruitment can be improved by providing benefits tailored to the needs of the target demographic, that is young people, and communicating that the benefits offered are not confined to the single dimension of monetary reward. The armed forces should not compete for people on the basis of pay, but should lever competitive advantage from its public service ethos and the benefits of the Service life. Life-satisfaction or work-life balance policies may be instrumental in attracting the current generation of prospective recruits who are less willing to accept the traditional demands of Service life[53] by illustrating the advantages that accompany the traditional restrictions.

AFOPS does not yet do this; it has merely established a change agenda[54] and provides a framework of HRM policies that can be taken forward by each of the Services. The armed forces need to recognise that they do more than provide an income. In doing so, they must do more than provide welfare support on a piecemeal basis but strategically integrate HRM policies to provide a life satisfaction for their most important asset: people.

REFERENCES

1. Hogarth T., Halsluck C. and Pierre G. with Winterbotham M. and Vivien D., *Work-Life Balance 2000* (DfEE Research Report No.249), *Labour Market Trends* (HMSO, July 2001).
2. Hodge M., Parliamentary Under-Secretary of State for Employment and Equal Opportunities quoted in 'Work-life Stress Costs Billions', *Financial Management,* Journal of the Chartered Institute of Management Accountants, June 2001, p.46.
3. Papalexandris N. and Kramar R., 'Flexible Working Patterns: Towards Reconciliation of Family and Work', *Employee Relations*, Vol.19 No.6 (MCB UP, 1997, pp.581–95.
4. Ministry of Defence, *Armed Forces Overarching Personnel Strategy*, MoD, London February 2000.
5. Guest D., 'Perspectives on the Study of Work-Life Balance', a discussion paper prepared for the 2001 ENOP Symposium, Paris, 29–31 March 2001, <www.ucm.es/info/Psyaplenop/grest.htm> accessed 14 February 2002.
6. Cully M., Woodland S., O'Reilly A. and Dix G., *Britain at Work: As Depicted by the 1998 Workplace Employee Relations Survey* (London: Routledge 1999).
7. Hogarth *et al.* (note 1).
8. TUC Website: <www.tuc.org.uk/work_life> accessed 31 January 2002.
9. DTI Website: <www.dti.gov.uk/work-life balance/case.ht> accessed 3 January 2002.
10. Pillinger J., Work/life Balance: Finding New Ways to Work, <www.tuc.org.uk/work_life/tuc-4022-f0.cfm> accessed 31 January 2002
11. The Future of Work Conference, Harrogate, June 2001, quoted in 'Women In Employment', *Labour Market Trends*, Vol.109, No.11 (November 2001) p.505.
12. Employees' Working Hours, *Labour Market Trends*, Vol.109, No.11 (November 2001) p.504.
13. Hodge (note 2) p.46.
14. Guest (note 5).
15. Maslach C. and Leiter M. P., *The Truth About Burnout* (San Francisco, CA: Jossey-Bass 1997).
16. Handy C., *The Hungry Spirit* (NY: Random House 1997) p.28.
17. Guest D and Conway N., *Fairness at Work and the Psychological Contract*, CIPD, quoted in Guest (note 5).
18. Cully *et al.* (note 6).
19. Hill E., Miller B., Weiner S. and Coliham J.,'Influences of the Virtual Office on

Aspects of Work and Work-Life Balance, *Personnel Psychology*, Vol.51 (Autumn 1998) pp. 667–83.

20. Burke R., 'Do Managerial Men Benefit From Organisational Values Supporting Work-Personal Life Balance?', *Women in Management Review*, Vol.15 No.2 (MCB UP 2000) pp.81–7.
21. Craig E. and Kimberley J., 'Mastering People Mangement: Work as a Life Experience', *Financial Times*, 5 November 2001, p.6.
22. Gratton L., 'A Real Step Change', *People Management*, 16 March 2000, p.28.
23. Tideman S., 'Gross National Happiness: Towards Buddhist Economics' adapted from a paper presented to a forum with leaders and scholars from Bhutan, in the Netherlands, January 2001.
24. Drucker P., 'They're Not Employees, They're People', *Harvard Business Review*, February 2002, p.76.
25. Ibid.
26. Guest (note 5).
27. Ministry of Defence, *The Strategic Defence Review: Supporting Essay 9, A Policy for People* (London: HMSO 1998) <www.mod.uk/issues/sdr/people_policy. htm>.
28. Ministry of Defence, *The Defence Mission* (London: HMSO 1998).
29. Bruce G., 'Armed Forces Overarching Personnel Strategy', Chapter 2 in Alexandrou A., Bartle R. and Holmes R. (eds.) *Human Resource Management in the British Armed Forces* (London and Portland, OR: Frank Cass 2001) p.14.
30. Ministry of Defence, *Armed Forces Overarching Personnel Strategy*, Chapter 3, Annex B, PSG 4 – Recruitment Policy, 2000.
31. *Armed Forces' Pay Review Body – 31st Report*, (Chair: Thorton-le-Fylde) London; TSO Cm536.
32. Ibid.
33. Spellar J. MP, Minister of State for Armed Forces, 'Implementing the SDR's Policy for People', *RUSI Journal*, Vol.144 No.6 (December 1999) p.63.
34. Alexandrou, Bartle and Holmes (note 29) p.2.
35. *Armed Forces' Pay Review Body – 31st Report* (note 31) pp.5–9.
36. Brightman B. and Moran J., 'Managing Organizational Priorities', *Career Development International*, Vol.6 No.5 (MCB UP 2001) pp.244–88.
37. Air Chief Marshal Sir Richard Johns quoted in Rogerson C. S. *An Analysis of the Cultural and Personnel Policy Changes Facing the Royal Air Force*, RMCS, No.11 MDA Dissertation, 1997, p.60.
38. Hunt J.W., *Managing People at Work*, 3rd edn. (Columbus, OH: McGraw-Hill 1992) p.43.
39. Guest (note 5).
40. Pledger M., DCDS(Personnel), 'Implementing a Policy for People', *RUSI Journal*, Vol.144 No.6 (December 1999) p.38.
41. Tonnies F. *Community and Society - Gemeinschaft und Gesellschaft*, translated and edited by Loomis C. P. (Ann Arbor: Michigan State UP 1957) pp.223–31.
42. Handy (note 16) Chapter 4.
43. Bett M. *Independent Review of the Armed Forces' Manpower, Career and Remuneration Structures, Managing People in Tomorrow's Armed Forces* (London: HMSO 1995).
44. Craig and Kimberley (note 21) p.6.
45. Wilkinson H., Working @ Life, *Management Today*, October 2000, p.36.
46. McEvoy A., 'Package Deal, Financial Management', *Journal of Chartered Institute of Management Accountants*, September 2001.
47. Guest D. quoted in Overall S., 'A Working Recipie for Quality of Life', *Financial Times*, 24 January 2002, p.13.
48. Handy C., *The Elephant and the Flea* (London: Hutchinson 2001).
49. Rogerson (note 37) p.66.

50. Hardman I. and Graham D., 'The Tyranny of Time: Balancing Work and Home in Dual-Career Households', paper presented by Department of International Studies, Nottingham Trent University at Gdansk, September 2001.
51. *Armed Forces' Pay Review Body – 31st Report* (note 31) pp.5–9.
52. Spellar (note 33) p.63.
53. Dandeker C., 'On the Need to Be Different: Military Uniqueness and Civil Military Relations in Modern Society', *RUSI Journal*, Vol.146, No.3 (June 2001) p.5.
54. Pledger (note 40) p.40.

9

Recruiting and Retaining the British Army Officer of the Future

RICHARD BARTLE

Cranfield University, Royal Military College of Science

Following on from a plethora of Human Resource Management initiatives in the British Armed Forces,[1] attention in the British Army has recently focused on the career management of soldiers and officers. The latest proposals for the Future Career Management of Officers[2] are designed to be part of the Army's Human Resource Management Strategy[3] and are intended to produce a system of career management that is more responsive to the needs of officers than at present.[4]

Central to this new career structure is the introduction of five 'fields of employment' (Defence Policy, Combat, Logistics, Human Resources and Technical) which, while giving officers the opportunity to specialise in certain areas, will retain enough flexibility in the system to ensure wide employability especially in the early stages of an officer's career. Ultimately, officers promoted to the highest levels will be involved in Higher Level Command and Strategic Management.

These major changes are being introduced in a continuing climate of recruitment and retention problems. It is among officers that difficulties with retention are most evident. Between 1 April 2001 and 1 December 2001 there was a net outflow of 260 officers – over 30 officers per month.[5] What is, perhaps, even more disturbing is that this haemorrhaging is most evident among the middle ranking officers (Majors and Captains) who would be expected to become the Generals of tomorrow.

In most civilian organisations this would not be a problem because 'people rise through an occupational career by changing firms and entering a given firm at an advanced level'.[6] The military, however, has a closed, internal market from which to recruit people for senior positions. They rarely recruit for the officer pool from outside. Thus any dilution in the numbers available for promotion could seriously impact on the ability of the Army to fill its senior posts.[7]

Various retention measures have been tried in the past, including loyalty bonuses and subsidised personal development courses[8] but the problem persists. It may well be that there are other, underlying reasons for the continued loss of officers at this crucial point in their careers that have more to do with the personalities of the recruits attracted to the Army and their eventual disaffection with a military career. Apocryphal stories tend to suggest that this has to do with the kind of jobs available to officers at the higher levels compared with earlier jobs in their careers. It could be that the Army recruiting strategy attracts personality types who prefer the action packed life of command at the junior levels over the increasingly business orientated desk-bound jobs of the higher levels of the military.

However, the new career structure for army officers may go some way towards remedying this providing the recruitment strategies and subsequent training and promotion systems allow for the different types of recruits that this scheme may attract.

Using research carried out in Britain and the USA, this chapter will examine the personality types of officers in the middle ranks of the British Army to see if any lessons can be learned for recruiting and retention strategies for the officers of the future.

PERSONALITY TYPE

Though not without its critics,[9] the Myers-Briggs model of personality type is, arguably, 'one of the most popular models of personality in the world'.[10] Isabel Myers and Katherine Briggs[11] based their theory of personality type upon Carl Jung's[12] work on different levels of consciousness and the influence of the conscious mind on behaviour.

Jung's first attempts to explain individual differences in personality centred around his observations that there were, fundamentally, two different types of people – 'extraverts' and 'introverts'.[13] He described 'extraverts' as being people whose energies are primarily directed towards their external environment of people and events; whereas, 'introverts' are those whose energies are directed inwardly towards thoughts and experiences in their inner environment. His further observations that there were still major variations in behaviour within the 'extravert'/'introvert' dichotomy led him to develop his theory of *Psychological Types*.[14]

'The essence of the theory is that much seemingly random variation in behaviour is actually quite orderly and consistent, being due to basic differences in the way individuals prefer to use their perception and judgement'.[15] Jung describes perception as the process of becoming aware of things, people, occurrences or ideas, and judgement as coming to conclusions about what has been perceived. He maintained that if people have different ways of 'looking' at things then they would inevitably have different ways of interpreting the information that they acquire.

Jung[16] went on to explain that there are two ways in which people perceive, either directly through their senses (sight, touch, smell, hearing and tasting) or indirectly into the unconscious mind through unconnected ideas, experiences and associations. He referred to the former as 'sensing' and to the latter as 'intuition'.

Similarly, he proposed two distinct ways of judging – 'feeling' which takes account of values and implications of decisions, and 'thinking' which is based upon logic and relies upon known facts.

Arguing that within the 'extravert'/'introvert' dichotomy there would be one dominant mental function – either, sensing, intuition, thinking or feeling – 'that is likely to be used most enthusiastically, most often and with the greatest confidence',[17] he posited eight personality types. Isabel Myers and Katherine Briggs[18] later extended this to the 16 types that are used today in the Myers-Briggs Type Indicator (MBTI).

The MBTI questionnaire was designed by Myers and Briggs to identify a subject's personality type and has been developed over more than 50 years into the present versions (which have some ethnic variations). The personality profiles produced by this tool are represented by four letters indicating the preferred poles of each of the dichotomies: Introvert/Extravert; Sensing/Intuition; Thinking/Feeling; and Judging/Perceiving (judging/perceiving relates to the ways in which individuals organise thought and decision making processes). (The author, for example, is **ENTP** – a preference for extraversion, intuition, thinking and perceiving.) The interpretation of the results from the MBTI is far from simple, consisting of three separate levels, each of which adds depth to the process by adding sub-scales and additional factors to the dichotomous preferences.[19]

It is beyond the scope of this chapter to explore these interpretations other than to mention that the results can be extremely useful in personal development, career and educational counselling, organisational development and psychotherapy. This chapter will concentrate on the uses of MBTI in career choices and how that might affect the recruitment and retention strategies for future army officers.

MBTI Types and Preference Groupings

There are a number of ways in which type preferences can be grouped together. The 16 types are usually displayed in a table that displays them in a way that makes interpretation easier (see Table 1).

TABLE 1
MBTI TYPE TABLE

	ST	SF	NF	NT
IJ	ISTJ	ISFJ	INFJ	INTJ
IP	ISTP	ISFP	INFP	INTP
EP	ESTP	ESFP	ENFP	ENTP
EJ	ESTJ	ESFJ	ENFJ	ENTJ

Source: Hirsh, S.K. and Kummerow, J.M. *Introduction to Type in Organisations*, 3rd edn. (Oxford: Oxford Psychologists Press 2000).

From this table a number of different groupings can be made which are helpful in describing various sorts of behaviour in the workplace. The most useful, when looking at career choice is to group the types in what is known as 'function pairs'[20] – **ST, SF, NF, NT** – which can be seen in Table 1 as the vertical columns in the MBTI Type table. Such pairs are often related to communication style, problem solving, career choice and organisational culture.

Table 2 is a chart of the Type Table Columns that shows the behavioural preferences common to each functional pairing. The distribution of these type pairings amongst British Army officers is somewhat unusual and will now be considered, initially, in relation to their American counterparts and, subsequently, with regard to both present and future career structures.

TABLE 2
TYPE TABLE COLUMNS

	ST	SF	NF	NT
People who prefer: Focus on:	Sensing + thinking What is	Sensing + Feeling What is	Intuition + feeling What could be	Intuition + thinking What could be
Contribute:	Polices and procedures	Customer service	Ideals worth striving for	Theoretical concepts
Have as a goal:	Efficiency	Helping others	Empowerment	Mastery
Ask questions about:	How will it be done, by when and for how much	Who will it affect who will do it, and how?	How will it be communicated and who will it impact?	What is the latest and most relevant theory?
	Bottom line	Offering support	Giving encouragement	Systems
Want teams to focus on: *May be found in*	Government, production, construction	Service, health care, education	Communication, arts, counselling and development	Technology, science, academia

Source: Hirsh S.K. and Kummerow J.M. *Introduction to Type In Organisations*, 3rd edn. (Oxford: Oxford Psychologists Press 2000).

Type Profiles of Army Officers

In 1996 Knowlton and McGee[21] published the results of their research into MBTI profiles of United States military officers. They used as their sample students from the Industrial College of the Armed Forces (ICAF). They argued that students of the ICAF are the equivalent of civilian managers in business and that it was from their ranks that virtually all of the strategic level leaders would emerge. Indeed, they stressed that, although not all the students of the college reach high rank (e.g. generals and admirals), almost all those who achieve high rank will be graduates of ICAF or similar Senior Service Colleges in the United States.

A similar situation appertains at the Royal Military College of Science (RMCS) in Britain. Students studying there for Master's degrees will, in the main, be Majors (or equivalent) and can thus be considered to be middle managers. Similarly, it is from their ranks that future strategic level leaders will be chosen. The importance of these groups to the military establishments of both countries is self-evident.

Knowlton and McGee[22] gave the MBTI questionnaire to the ICAF classes of 1994 and 1995. At the RMCS, the author gave the questionnaire to several Master's degree groups during the period 2000 to 2002 (n=239). The only difference between the two samples being that those from the RMCS

TABLE 3
MBTI RESULTS – RMCS (n = 239)

ISTJ	ISFJ	INFJ	INTJ
17.9%	1.6%	0.8%	8.7%
ISTP	ISFP	INFP	INTP
3.2%	0%	2.1%	3.7%
ESTP	ESFP	ENFP	ENTP
6.2%	2.5%	2.1%	9.6%
ESTJ	ESFJ	ENFJ	ENTJ
20.5%	3.7%	2.5%	14.2%

were all army officers. Table 3 above shows the MBTI profiles for the army officers at RMCS:

Some interesting observations can be made for these

results especially the preponderance of ST and NT types compared with the SF and NF types. ENTPs and ENFPs in total only represent 11.7 per cent of the population at RMCS

TABLE 4
DISTRIBUTION OF TYPES – ICAF/RMCS

	ST	SF	NF	NT
ICAF	51	8	9	32
	ST	SF	NF	NT
RMCS	48	8	8	36

Adapted from: Knowlton B. and McGee M. *Strategic Leadership and Personality: Making the MBTI Relevant*, 2nd edn. (Washington DC: National Defense University, Industrial College of the Armed Forces 1996).

and, so far, there has been no one with a type ISFP preference. Table 4 shows the comparison between ST, SF, NF and NT results from both ICAF and RMCS.

What is most striking here is the similarity between the two sets of results which, statistically, are not significantly different ($X^2(3)= 0.953$). But the RMCS results are quite different when they are looked at in the context of the distribution of types within both the UK population as a whole and the UK male population. These are shown in Tables 5 and 6 opposite.

The first, most apparent difference, is that between the **SF** samples for both of the UK populations and that of the UK military. Only 8 per cent of the UK military falls into the **SF** category whilst the figures for the population as a whole and the male population are 40 per cent and 23 per cent respectively.

Furthermore, the **ST** and **NT** results for the RMCS account for over 80 per cent of the cohort whilst those for the whole UK population and for UK males are 46 per cent and 65 per cent respectively. In addition, more than twice the RMCS cohort (36 per cent) are **NT**, compared with only 10 per cent of the UK population and 15 per cent of UK males.

As might be expected from these results the average profiles for each group is different. The UK officer profile is **ESTJ** whereas the UK population profile is **ESFJ** and the male

TABLE 5
DISTRIBUTION OF TYPES – UK POPULATION (n=1634)

ISTJ 13.7%	ISFJ 12.7%	INFJ 1.7%	INTJ 1.4%
ISTP 6.4%	ISFP 6.1%	INFP 3.2%	INTP 2.4%
ESTP 5.8%	ESFP 8.7%	ENFP 6.3%	ENTP 2.8%
ESTJ 10.4%	ESFJ 12.6%	ENFJ 2.8%	ENTJ 2.9%

ST 36%	SF 40%	NF 14%	NT 10%

Source: Kendall E. *MBTI Step 1 European English Manual Supplement* (Oxford: Oxford Psychologists Press 1998.

TABLE 6
DISTRIBUTION OF TYPES – UK MALES (n=748)

ISTJ 19.7%	ISFJ 7%	INFJ 1.6%	INTJ 2.5%
ISTP 10.8%	ISFP 3.7%	INFP 3.6%	INTP 4.1%
ESTP 8.2%	ESFP 6.1%	ENFP 5.1%	ENTP 3.6%
ESTJ 11.6%	ESFJ 6%	ENFJ 2%	ENTJ 4.3%

ST 50%	SF 23%	NF 12%	NT 15%

Source: Adapted from Kendall E. (see Table 5).

profile is **ISTJ**. These are curious results given that the work of Kroeger and Thuesen[23] suggests that the personality type most attracted to the military life is **ISTJ**, which is the average profile for the UK male population, while the average for the RMCS cohort is **ESTJ**. This might say something about the lack of success of the recruiting strategy for the British Army.

In addition, Knowlton and McGee[24] have suggested that the best personality types for strategic leadership at the highest level could be **ENFP** and **ENTP**. Though the Army profile is not a fit with either of these personality type profiles, the British Army might take some slight consolation in that their combined occurrence is slightly higher in the military profile (11.7 per cent) than that of either the total UK population (9.1 per cent) or the male population (9.7 per cent). However, a further comparison might throw some more light on the relationship between the RMCS army officers' profile and that for strategic leadership

As well as positing a profile for strategic executives, Knowlton and McGee[25] produced tables showing the profiles for three tiers of management: supervisors, managers and executives (strategic managers). They made comparisons between the profiles of the strategic leaders and those of the ICAF students. Given the similarities between the ICAF and the RMCS profiles it may be argued that further comparisons are superfluous and that the ICAF comparisons would be enough to prove a point. Nevertheless, for the record, a comparison of ICAF students, RMCS army officers and

TABLE 7
COMPARISON OF BUSINESS EXECUTIVES WITH THE ICAF
AND RMCS STUDENT SAMPLES

	ST	SF	NF	NT
Executives	30%	3%	15%	52%
ICAF	51%	8%	9%	32%
RMCS	48%	8%	7%	36%

Source: Adapted from Knowlton and McGee (note 21).

business executives is produced in Table 7.

The differences between both the ICAF and RMCS samples and those of the business executives are statistically significant ($X^2(3)=16.625$ and ($X^2(3)=11.928$ respectively). There are also some interesting comparisons to be made, for example, the ICAF/RMCS **ST/SF** populations are almost twice the size of the business executive sample and there are 20 per cent more **NT**s in the business executive sample than in that for ICAF/RMCS.

In addition, while the ICAF population is generally **ISTJ** and the RMCS officer population is generally **ESTJ**, the executive manager profile is generally **ENTP**. This **ENTP** profile is the same as that which was surmised by Knowlton and McGee[26] to be one of the two ideal profiles for strategic managers, the other being **ENFP**.

At the present, therefore, the Army does not seem to be either recruiting or retaining sufficient officers of the necessary personality type most suited to strategic leadership. The question remains as to whether their present recruiting strategies will provide them with enough officers to fill all the new positions created by their plans for five fields of employment for future officers.

PLANS FOR 'THE FUTURE CAREER MANAGEMENT OF OFFICERS'

Although the published plans for officers' careers[27] are clear that most officers may serve in a number of fields, the general intent of the paper seems to be that, at the later stages of an officer's career, he/she is likely to concentrate on one area of expertise. It would also seem to be the case that it is hoped that the attraction of concentrating on one of these areas of expertise will both recruit and retain officers of the future. It would, therefore, seem appropriate if at this stage these five fields of employment and the personality types that are likely to be most attracted to each of these fields were examined more closely.

Exact details of each of the five fields of employment have yet to be published but Table 8 gives an outline description of each field.

TABLE 8
ARMY OFFICER EMPLOYMENT FIELDS[28]

Employment Fields

Combat	Posts with a significant operational role
Human Resources	Posts responsible for the development of individuals
Defence Policy	Posts involved in the formulation of defence policy
Technical	Posts involved with systems and/or equipment
Logistics	Posts involved in logistics management

Combat Role

It has been argued that it is this role that presently attracts most officers to the Army.[29] There is a small difference between the ICAF officer group type profile (**ISTJ**), and that of the RMCS officers (**ESTJ**); however, they are both markedly different from the suggested and actual profiles for strategic managers (**ENTP** and **ENFP**). It is possible that the **ISTJ/ESTJ** profiles represent those types of personality that are attracted to the idea of leading men and women in the field, that is combat leadership. This kind of activity is mostly carried out by the lower officer ranks up to and including Major.

Beyond this rank the emphasis moves more and more towards strategic leadership. Yet this seems unlikely to be what the **ISTJ/ESTJ** types have joined for. Furthermore combat leadership has been, hitherto, the basis upon which they have been promoted.[30] As so many officers are leaving the army at the Captain/Major level,[31] it may be due to the recognition that they are no longer suited to the kind of jobs available to them above these ranks. They may also realise that promotion to the higher ranks may be more difficult given their own personality types and the new requirements of these desk jobs for skills in strategic leadership. So, while the combat career seems to attract most officers, it also seems to be the one which loses most officers at the Major stage. Perhaps a new approach to training that takes into consideration the personality types of these officers might result in greater retention.

Human Resources

The short outline above describes this field as involving the development of people. By that one can assume that it is to do with education and training. In Table 2, it was suggested that the personality types usually found in education and development were the function pairs **SF** and **NF**. **SF** is not only the most *under-represented* function pairing in army officers (8 per cent) it is also the most *represented* in the UK population as a whole (40 per cent). So it would seem that the army's recruiting system is failing miserably as far as 40 per cent of the population is concerned. In addition, the **NF** pairing is also under represented (8 per cent in the Army and 14 per cent in the UK population).

However, it could be argued that it is not the recruitment strategy per se that is failing but the selection system which has been said to be self-perpetuating. That is to say, the army officer selectors recruit in their own image and reject those who do not fit into their perception of the traditional army officer.

Furthermore, the promotion system, particularly at the lower levels, may well be exacerbating this problem by not promoting those with the **SF/NF** personality profiles because they do not fit into the role of combat leader. However, whether it is the recruitment strategy, the selection system or problems with the promoting **SF/NF**s it seems self-evident that there is a problem for the Army in attracting and retaining people for this employment field.

Defence Policy

It has already been pointed out that there is a paucity of officers with suitable type profiles for strategic management (**ENTP** and **ENFP**). Yet it seems reasonable to surmise that it is strategic managers who are needed to be 'involved in the formulation of defence policy'. Unfortunately for the Army, people with these type profiles are rare. To some extent the Army has already recognised the need for strategic management skills and it has been at the forefront in introducing a Master's degree in Defence Administration (MDA) for its middle ranking officers. However, most officer leadership training is combat orientated. What does not seem to have been considered is the need for some form of leadership training that will result in the kind of 'personal integration and transformation' that can produce 'outstanding leaders'.[32]

Such training would recognise the large number of officers who have personality preferences that make it difficult for them to perform to the best of their ability in strategic leadership roles. For, the conscious development of the non-preferred functions *is* possible and people can be trained or counselled to further this development.[33]

Consequently, it may be that the Army can achieve both improved retention and better equipped strategic leaders by designing new leadership training that will develop the non-preferred functions of those predominantly **ESTJ** officers who make up such a large percentage of junior and middle ranking officers in the British Army. These officers may then, perhaps, be more inclined to stay in the Army because they will be better prepared for the difficult task of strategic leadership.

Technical

Fortunately for the Army, in this employment field there would seem to be a higher percentage of officers (36 per cent) with the preference grouping most associated with technological and scientific work (**NT**) than can be found in the general population (10 per cent). It could be assumed, therefore, that this bodes well for the Army in this particular field. However, it must be remembered that the new career structure is intended to retain enough flexibility in the system to ensure wide employability especially in the early stages of an officer's career. Thus, while being suited to the technological field it does not follow that officers with this personality type will be at all suited to working in other fields such as Human Resource Management or Defence Policy. Consequently, although officers of this type may be extremely motivated to stay in the Army whilst in the technological field, it may be that they will be less motivated (even to the extent that they may leave) when posted into other employment fields.

Furthermore, there still remains the problem of strategic leadership at the higher level and whether these officers with personality types best suited to technological careers will be suited to working outside their area of preference.

Logistics

It could be argued that logistics is the career field that is best represented in the type profiles of army officers. Indeed, 48 per cent of army officers are to be found in the **ST** pairing (which shows a preference for working in production and construction). Even so, this is smaller than the representative percentage of all UK males (50 per cent). However, the areas said to be of interest to **ST**s concern asking the questions – Although these questions are obviously important in logistics, perhaps the questions – Who will it affect, who will do it and how? – are also just as relevant. These are said to be the kinds of questions likely to be asked by people who fall into the **SF** pairing – the least represented function pair amongst army officers.

CONCLUSIONS

This chapter began by summarising some of the problems facing the British Army with regard to its officer recruitment and retention and describing the new career structure that is designed to alleviate these problems. While the results of the research carried out for this chapter may not offer a panacea for these problems they do raise some questions about them and also about the training of officers for higher command.

Arguably the most intriguing aspect of the results is the predominance of **ST** and **NT** types in the RMCS sample. Reference to Table 2 shows that this representative sample of army officers at RMCS are, predominantly, suited for careers in government, production and construction (**ST**) or technology, science and academia (**NT**). There is a distinct lack of **SF** and **NF** types who are said to prefer careers in service, health care and education (**SF**) and communication, counselling and development (**NF**). Yet it has been shown that people with these types of personality will definitely have a place in the five career fields of tomorrow's army.

However, for whatever reason, these people are not being attracted to a career as an army officer. This must be a matter of great concern for the Army. Their new officer career structure hangs on attracting the right people to fill the five new career field posts and then subsequently retaining them. At the present time this is obviously not happening, the Army recruiting and selection system is failing to attract people with **SF** and **NF** personality types.

Whether people will be attracted to any of these careers in the future will depend upon what new recruiting, promotion and training strategies are introduced to underpin the new plans for future career management. Certainly something has got to change; to stay with the old methods can only be a recipe for decline not rejuvenation.

REFERENCES

1. For example see: Strategic Defence Review (1998), the launch of the Tri-Service Equal Opportunities Training Centre in 1998, the MoD's Mission Statement (1998), the 1999 Defence White Paper, the Social Code of Conduct (2000) and the Armed Forces Overarching Personnel Strategy (2000).
2. APC, *The Future Career Management of Officers*. ECAB/P(00)02, 9 March 2000.
3. ECAB, *The Army's Human Resource Management Strategy*. ECAB/P(97), 10 July 1997.
4. APC, *The Future Career Management of Officers* (note 2).
5. Ministry of Defence, *Defence National Statistics: Armed Forces Personnel Statistics*. Defence Analytical Statistics Agency Website: <www.dasa.mod.uk/servstats> accessed Jun 2002.
6. RAND, *Future Career Systems for US Military Officers* (National Defense Research Institute, Defense Manpower Research Center, RAND 1994).
7. Dandeker, C. and Paton, F. *The Military and Social Change: A Personnel Strategy for the British Armed Forces* (London: Brassey's for the Centre for Defence Studies 1997).
8. Bartle, R.A. 'Continuous Personal Development in the British Army: A Recipe for Retention or a Motive to Move?' in Keri Spooner and Colin Innes (eds.) *Employment Relations in the New Economy: Proceedings of the Ninth Annual Conference of the International Employment Relations Association*, 2001, Vol.1, pp.1–12.
9. See, for example: Zemke, R.. 'Second Thoughts About the MBTI', *Training*, Vol.29, Issue 4 (1992); Keirsey, D. *Please Understand Me II* (Del Mar, CA: Prometheus Nemesis Books 1998); Carrol, R.T. *The Skeptic's Dictionary* (Hoboken, NJ: John Wiley 2002) <http://skepdic.com/myersb.html>.
10. Teamtechnology, 'Working out your Myers-Briggs type', 2001. <www.teamtechnology.co.uk>.
11. Myers, I.B. with Myers P. B. *Gifts Differing: Understanding Personality Type* (Palo Alto, CA: Davies-Black 1995).
12. Jung, C.C. *Modern Man in Search of a Soul* (New York: Harvest Books; Harcourt, Brace, Jovanovich, 1933); and Jung, C.C. *Psychological Types* (Princeton UP 1976).
13. Myers, I.B. *An Introduction to Type. A Guide to Understanding Your Results on the Myers Briggs Type Indicator* (Oxford: Oxford Psychologists Press 1998).
14. Jung, C.C. *Psychological Types* (Princeton UP 1976).
15. Myers (note 13) p.6.
16. Jung 1933 (note 12).
17. Myers 1998 (note 13) p.22
18. Myers with Myers (note 11).
19. Ibid. p.10
20. Hirsh S.K. and Kummerow J.M. *Introduction to Type in Organisations*, 3rd edn. (Oxford: Oxford Psychologists Press 2000) p.7.
21. Knowlton, B. and McGee, M. *Strategic Leadership and Personality: Making the MBTI Relevant*, 2nd ed. (Washington DC: National Defense University, Industrial College of the Armed Forces, 1996).
22. Ibid.
23. Kroeger O. and Theusen J.M. *Type Talk* (New York, NY: Dell 1988).
24. Knowlton and McGee (note 21).
25. Ibid. p.34.
26. Ibid.
27. APC, *The Future Career Management of Officers* (note 2).
28. Army Board, *Officer Career Development*, July 2001.
29. Bartle, R.A. 'The Use of Personality Type in Solving Recruitment and

Retention Problems in the British Army', paper presented at the 10th IERA Conference, Broadbeach, Australia, July 2002.

30. Garnett, J.W. 'The Future Development and Selection of Junior Commanders', unpublished MDA Dissertation, RMCS, Cranfield University, 2000.

31. Ministry of Defence, *Defence National Statistics: Armed Forces Personnel Statistics* (note 5).

32. Ingalis, J.D. 'Genuine and Counterfeit Leadership: The Root Causes of Human Capital Flight', *Strategy and Leadership* Vol.20, No.6 (2000) pp.16–22.

33. Myers, D.K. and Kirby, L.K. *Introduction to Type Dynamics and Development*, European English Edition (Oxford: Oxford Psychologists Press 2000).

10

Infantrywomen – An Ethical Dilemma?

GEORGINA NATZIO

Does the idea of employing infantrywomen pose an ethical dilemma? What follows here is an attempt to find out. Evidence from World War II and after, shows an ethical and practical fluctuation in the whole area of female military employment. In this chapter an attempt is made to try and discover why such fluctuation occurred, and why the idea of a military life for women, whether or not in infantry units, has persisted in our culture.

Ethics and national interests do not always coincide, but the British have a long history of striving to bring the two together in the process of governing, not only their overseas territories but also the homeland and its great institutions. Study of British war management during World War II reveals that tradition was alive and well at the time. British homeland defence experience reviewed from War Cabinet papers, War Office and other military archives, together with historical accounts of the defence of the Soviet Union, also suggest that in certain circumstances cherished or fiercely-held ethical attitudes may alter, be traded off one against another, or be put into temporary suspension to suit a wider national interest, particularly when dictated by the demands of survival.

With his answer to a written question from Member of Parliament, on 22 May 2002, (accompanied by supporting material[1]), the British Secretary of State for Defence, Geoff Hoon, announced his decision against including women in infantry units. He may not have known it but in arriving at his conclusion, he was following a precedent for restraint set by the War Cabinet during World War II, as its members evolved their manpower-distribution policy. Commonly-perceived today as radical, especially when it came to female employment, in practice that decision was largely shaped by a conservationist approach. Memories of World War I losses incurred little more than 20 years earlier may well have still resonated among them – scarcely a family in the country had remained unscathed.

Returning to the present, 'The key issue', Mr Hoon wrote, 'was whether the inclusion of women in close combat teams could adversely affect the combat effectiveness in those teams in a high-intensity direct fire battle.' All the units under review for female inclusion operated primarily in small units as fire teams, or tank crews.

The evidence available, he added, suggested that on operations other than close combat, the presence of women in small units did not affect performance detrimentally. However, there was no evidence to show whether this remained the case under the extraordinary conditions of high-intensity close combat. Lacking relevant direct evidence from either field studies or other countries' experience, he had decided to rely on military judgement to form the basis of his decision. Military advice was that in such fighting units, group cohesion became of much greater significance to team performance and that the consequences of failure could have far reaching and grave consequences. To admit women would involve risk without any offsetting gains in terms of combat effectiveness.

These remarks deserve interpretation.

First, they meant that female presence would interrupt the critical processes of infantry group-bonding – ancient methods by which such warrior-groups organise themselves, by formally and informally allocating overt, or hidden roles to individuals who reveal the best aptitudes.[2] This is intricately-connected to questions of survival.

Second, it will also be remembered that there had been widespread concern when the idea of infantrywomen was being mooted during the 1990s, about male-female strength-differentials, a crucial measurement of effectiveness.

Ancient means of weighing military potential, by trainers and among comrades, have always been based on unearthing qualities of character and physique which would enable a man ultimately to stand and fight. Elements of this remained true in the United Kingdom's large volunteer, then conscript, armies of the 1940s even though professional military judgement had to be augmented by medical support, to cope with the numbers entering basic training. A requirement for both traditional and medical assessment of recruits intensified, however, in the training of the steadily-reducing, highly-specialised, and eventually volunteer British forces typically employed on national business after the Korean War.

The higher the degree of combativeness and endurance required in more specialist units, the heavier the demand on the individual. Forget, however, the idea that hand-to-hand encounters in war since 1938 have been increasingly rare (as far as this writer has learned to understand it) for that is not the point at issue. By what some might consider to be the best standards of military practice, close-quarters combat is never to be ruled out as a possibility and forms an intrinsic part of the mosaic of the unexpected which recruits have traditionally been taught to understand. '*Mais*', in the immortal words of Paris-born Alphonse Karr (1808–90), novelist, critic and sometime Editor of *Le Figaro: 'Plus ça change, plus c'est la même chose'*. But the more things change, the more they remain the same.

Asymmetry, a term in currency in the USA by 1997, though, has existed ever since earliest developments in military technologies – as uneven, perhaps, in the early history of Chinese (Gunpowder) or Ancient Greek warfare (simple catapult, torsion catapult), for instance, as they are now – when one side has a grasp of new equipment or stratagems initially foreign to the other. Therefore, in order to repose complete confidence in new technologies regarding their influence on the type of warfighting, even as adaptation to their use takes place, there has to be room for a preparation for collapse, which itself stimulates tactical, if not strategic ideas for determined reversal of bad fortune. There has to be a clear vision of potential disaster where, if time and circumstance allow, hand-to-hand struggle may be needed in order to achieve an objective; and where individual resourcefulness and stoicism, altruism even, under profoundly-violent, novel and personally highly-dangerous conditions, will be exacted as the price of success against all odds.

This represents war's true reality – instinctively under-stood by most men, especially those with a talent for war, but as yet, few modern women, save perhaps for those with experience from serving in the 1991 Gulf War, or who have careers or other associations with military and defence communities.

THE TWICE-BITTEN FACTOR

Despite some opposition, which had its roots in perceptions of the actuality of war – among senior naval officers for example – women nevertheless continued to be recruited for the three British armed services post-1945. There was, though, considerable reduction in female personnel strengths after demobilisation and heavy reduction in their scope of employment after 1949 – when the Women's Services were put onto a permanent footing. Regardless of British involvement in a continuum of restricted external conflicts from that year on, female resumption of some old tasks and extension into training for new occurred only slowly in conditions of relative world peace between 1970 and the year of writing, 2002.

Because of that retrenchment between 1945 and the 1970s, World War II therefore remains the main source of knowledge of how British women, as military auxiliaries, stood up to the pressures of real war, though banned from direction to the teeth arms, and, among other prohibitions, from carrying personal weapons – until 1980–81. The arming picture is confused, with each of the Women's Services taking up training at its own pace.

A curious anomaly here is that many were doubtful about increasing women's proximity to the fighting units, yet those who were recruited, either as volunteers or conscripted during the 1940s to combatant jobs (short of armoured or infantry) were by rote of another attitude officially-prohibited from carrying arms, even for self-defence. In fact, photographic and anecdotal evidence reveals that weapons-training for the auxiliaries was carried out, and that weapons were issued when it was considered necessary by units taking their own steps. Members of the Women's Auxiliary Air Force (WAAF) were given rifle and hand-gun training during the Dunkirk crisis, yet one report has it that no official weapons-training was given to Auxiliary Transport Service (ATS) members who volunteered to join the mixed batteries destined for the Continent in 1944.[3] ATS serving with the anti-aircraft (AA) batteries were not officially allowed to fire the guns, and heavy work on the gunsites was also prohibited.

Again, anecdotal evidence of varying quality suggests that both these latter bans were also broken from time to time. While that particular military or humane injustice regarding weapons-training has been officially-removed, the attitudes underlying it still appear to exist with regard to female infantry-training. Genuine reluctance to see further militarisation of our society must be honourable, yet how can this be squared with present and possibly future demands which may be made on our armed servicewomen?

In our own era, driven by apparently ever-diminishing resources in personnel as well as finances, as much as by non-discriminatory and human rights legislation, we are still thinking through how military employment might most effectively be distributed among male and female recruits, as General Sir Frederick Pile (GOC-in-C, Anti-Aircraft Command) a huge employer of female auxiliaries, did. He also faced cash problems and personnel shortages on a huge scale from 1938 onwards (demands for trained men from Anti-Aircraft Command were running at the order of 19,000-odd at one time). The British experience of pressure to make intelligent use of individual resources during 1938–45 showed, for example, that the introduction of female personnel was directly connected with the processes of managing technological change, as well as the problems of manpower shortage.

Ergonomics as a subject was defined then in relation to matters arising from the man-machine interface and today is understood as the science of making the job or equipment fit the worker – based on the study of individuals in their working environments. It was a discipline which was not quite invented in World War II, but as knowledge and understanding of the subject's importance became more widely appreciated, it has become securely lodged in the annals of military practice. Thousands of women released after basic and then specialist training into jobs with the three armed services, especially those thousands recruited for the mixed AA batteries, drove the development of military science in this area. It also came to be applied to male volunteers and conscripts – as did a very great interest in the search for further means of avoiding breakdown from exhaustion, for instance, by both sexes; a search itself both humanitarian and pragmatic.[4]

For those working in the wartime civil and male armed services, together with senior women officers from the WRNS (Women's Royal Naval Service), ATS and WAAF, who between them dragged order out of a highly-confused manpower situation and survived, at times, a phenomenally-bad press between 1939 and 1943, ethical attitudes also played a large part in their decision-making. They directly-influenced the way they approached a vast number of problems, social and organisational, which they had to overcome as their organisations were set up and then expanded. Ethical thought was applied to such matters as shift lengths, accommodation, codes of discipline and behaviour; health, the status under international law of WRNS codes and cipher specialists serving aboard convoy ships.

There was a double edge to this exercise across a broad range of ethical practice, for the maintenance of public morale was intricately involved in the achievement of high standards not only of professional training for the auxiliaries but also in their care and management. This was buttressed by devising, from the ground up, disciplinary codes especially for the female auxiliaries. It was not only the sensitiveness of the three Commandants, and their staffs in understanding the special requirements and attitudes of women, then, which needed to be reconciled with those of the male organisations, but the hard and sustained work the women themselves put in to earn the respect they eventually won from their male peers in their host-organisations, and approval and acceptance from the public.

The establishment of good, solid, reciprocal working relationships with their male counterparts was a collective female achievement carved out of a hostile environment in every sense of the word. Upon this was placed the weight of public confidence and goodwill which directly, in turn, affected the amount of support the incoming girls and women received from the communities where, for example, their guns were sited.

Some of the ground-work for this fundamental attitude towards women in the RAF, which might perhaps still be detected in male-female working relationships in the service, was laid by individuals whose approach reflected that of Air Marshal E. L. Gooaage, AOC Balloon Command. The Air Marshal oversaw assessments made by his subordinates and medical officers of airwomen's fitness for joining balloon crews. Originally doubtful, Gossage was scrupulously fair-minded and his extensive report reflects his and his staff's considered and thoughtful appraisal of their experiment, meticulously documenting their change of mind.[5]

A description of his personality[6] further suggests that, nevertheless, if the Air Marshal had considered airwomen could not carry out the balloon crew jobs, or were receiving injuries at a rate he found unacceptable, he would have taken steps to ensure that their employment in balloon crews would not have gone ahead.

This humane, yet pragmatic approach, might well also have been applied to male balloon crews – even in an era when men were automatically expected to put up with considerable hardship and when resolving unacceptable injury levels from equipment would have been far harder for *them*. Improvements in balloon-handling equipment and techniques made in 1940, however, initially helped the airmen and, then, to make the experiment of employing female crews a success. Gossage decided to opt for 100 per cent female crews as he 'foresaw administrative difficulties with mixed crews', as he put it. The women's health was scrupulously watched and accounted for against other trades and women's health in civilian work.

A minimum height of 62 inches for the first intake was lowered to 60 inches and applicants had to be able to lift heavy weights, minimum 40lb. By the time Gossage's report was made in September 1942, after the official launching of the scheme and passing out of the first trainees on 10 July 1941, over 16,000 airwomen had met those fitness standards. Only three serious injuries had been recorded thus far. Serious injuries did occur later on, however – whiplash was a hazard when cables were severed, during bombing raids or if enemy aircraft flew into them.

The exertion of steady goodwill and good sense, therefore,

overcame pockets of male resistance which engendered female defensiveness and instead promoted real mutual trust and respect. Shared danger, already mentioned, was a powerful impetus for concentration on mutual efficiency. Initially dismayed by the introduction of women to Balloon Command to replace them, airmen, when they fully understood that they were urgently needed elsewhere, took on the training of airwomen with enthusiasm. 'It was pleasant also to watch the scepticism – almost suspicion – of officers concerning the experiment disappear as substitution progressed, by reason of the keenness and ability displayed by the airwomen', Gossage wrote.

All that has been written thus far is not intended to be a sentimental historical assessment. For the organisation of Britain's defences against air attack (Air Defence of Great Britain or ADGB) to work, cooperation between Army and RAF, and between female and male organisations had to be practised at a high, not to say intricate technical and personal level and sustained over years. The male achievement was quite equal to that of the women, for like them, the men were also required to make a huge leap of the imagination. It was required of the men to understand that their own effectiveness was not in doubt because women had been recruited to augment, alongside them, their efforts to repel enemy assault. The women's reward was genuine male appreciation for the high levels of skills and expertise they came to acquire in unfamiliar, sometimes alien employments; the men's to find new, unconventional, comrades on whom they could rely.

ETIIICS V. SURVIVAL?

British military archives from both World Wars, therefore, collectively demonstrate in different forms, the existence of a constant tension between ethics and pragmatism in military practice and democratic wartime government. Concerning the employment of women as substitutes for men virtually throughout the infrastructure of all three armed services, hard questions were jointly tackled by politicians, military and civil servants, men and women, alike. Surviving records are remarkably frank. Disagreeable discussions occasionally took place which make unpleasant reading now. Difficult and delicate subjects were faced by both sexes unaccustomed to the levels of openness required and for whom there was little, either in language or custom to help them.

In deference to those sustained efforts we can do no less. Due to what some might regard as the implacable demands of war, or potential war, we find ourselves in a position now, in spite of all that has gone before, therefore, to have to ask:

'Is ethical conflict built in to the employment of women as infantry?'

'Should women, pan-culturally symbols and promoters of peace and civilisation (albeit with some notable warrior exceptions) yet still principal life-carriers and nurturers, be ordered in the national interest to take life?'

'Should they be employed in the teeth arms during relative peace, rather than great national danger?'

'With little cultural knowledge of, or instinct for the demands warfare makes on infantry, can they be regarded as reasonable candidates for front-line regiments?'

By relaxing all remaining restrictions on women's military employments would we be accepting the possibility of annihilation, something the War Cabinet refused to do during 1939–45 judging, that is, by the limitations set over which tasks should be made over to the WRNS, the ATS and WAAF?

MILITARY ETHICS – LUXURY OR NECESSITY?

The ethics of employing female infantry became a luxury the Soviets could not afford to deliberate over for long during World War II. After the German invasion in 1941 so steep were the manpower losses that by 1943 women of the USSR had entered all the armed services. They could be found in infantry, anti-aircraft defence, tanks, transport, communications, air- and partisan-warfare and medicine. About 500,000 women are estimated to have served at the front, either in combat, or support. Greisse and Stites noted in their much relied on account that the Komsomol (Communist Party Youth) schools trained about 250,000 as mortar-women, heavy and light machine-gunners, automatic riflewomen, snipers and riflewomen. Theirs and other investigations estimate that about one million Soviet women, to 12 million men actually fought.[7] Antony Beevor's description of the bitter contest for Stalingrad has some valuable material about Red Army women's contributions, aligned with contemporary reports from the German Army.[8]

According to Richard Overy, 23 of the 46th Guards Women's Night Light-Bomber Regiment became Heroes of the Soviet Union. By 1945, there were 246,000 women in uniform at the front. The total employment figure, by comparison, for the officially-unarmed ATS in UK and abroad, as at 30 September 1943 was 214,420. However Overy goes on to point out that, unlike the British experience, 'of the 34.5 million [Russian] men and women mobilized an incredible 84 per cent were killed, wounded or captured'.[9] Total military deaths from all causes amounted to 8.6 million, which he regarded as the most reliable figure. Other revised estimates produced by Russian historians, he added,[10] suggested an even higher figure but there were complications over the computations.[11]

Numbers, therefore, gained an untoward prominence in those World Wars. As any student of war will know, however, numbers became an increasingly vexed question as that century wore on . In order to understand quite how savage personnel cuts were perceived to be at a deep level, even among generations who had not experienced large-scale conflicts, we have to take a brief look at the ancient origins of numbers dependency. Herein lies the other half of the dilemma under investigation.

In early contests for territory in China, for example, large armies became increasingly sought, even before the acquisition of more expensive equipment, in the pressure for a successful result. As Sun Tzu wrote: '1. Generally, operations of war require one thousand fast four-horse chariots, one thousand four-horse wagons covered in leather and one hundred thousand mailed troops.'[12] Although armies of such size were unknown in China before 500 BC, Griffith notes in his translation of 1963 that the author of *The Art of War*, on the other hand lived at a time when large armies were effectively organised, well trained and commanded by professional generals. Can it be inferred from this, that the greater the territorial gain, the higher the professionalism, the greater the dependence on numbers?

A visceral dependence on numbers in our own times might well be allied to increased lethality of military technologies, accompanied by fine appreciation of the value of comprehensive and efficient organisation. If true, this would shed some light on reasons for the persistence of women's presence in all the armed forces of the NATO Allies as presently constructed. A contradiction is created on the one hand by some weapons systems' high lethality putting pressure on manpower survivability, against a background of intense establishment reductions – this last being paradoxically reinforced by other systems which reduce the requirement for operators.

That labour-saving element in improvements in military technology do not progress in tandem with developments in the nature of warfare itself, comes as no surprise and the mechanics of this particular asymmetry may themselves form an underlying reason for persistent reliance on known factors that have brought success in the past. With clear, traceable links for the British back to 1917, that is, to desperate experience of the huge demands on supplies of men, of total war, numbers dependency may be a difficult, if not impossible habit-of-mind to shrug off.

The paradox is that the numbers instinct co-exists with another, equally profound, eloquently described by Sir John Keegan, which also derives from a desire to survive. This could be outlined as an urge to prevent women from being killed or captured, and from fighting to the death, as the prospects for a future would be thus become seriously questionable. The treatment of women by conquering armies in our modern era since 1945 lends strength to the rationality and depth of that perception. What happened by 1917 and again by 1942, however, was that in real war conditions the numbers-reliance instinct was balanced by a profound need to retain a vision of a real future. A wish surfaced surprisingly early during World War II to prepare for the future . It recurs repeatedly throughout the armed services' archival material written by men, as well as by women who might be expected to carry an instinctive sense of it.

To sum up, one of the reasons for continuing to recruit women into military units since 1945 and, again, after the ending of our national service in the 1960s, has been a much

quoted shortage of young men of the right combatant age-
groups. The American position was memorably set out by
Binkin and Bach, for the Brookings Institution, in their survey
Women in the Military. With the decision to end the draft, in
1970, the United States embarked on a venture unprecedented in any nation's history: to field a force of over two
million relying solely on volunteers. Could enough men be
found, willing and able to volunteer, without exorbitant
additional costs and without compromising the quality of
military manpower? Reasonable people disagreed: 'Although
it appreciated the great uncertainties involved, the
Department of Defense realised it had to further expand the
role of women.'[13]

The atmosphere was, however, considerably soured by
litigation driven by feminist pressure, to outlaw what were
regarded as discriminatory military practices during the
processes of expanding female military employment in the
US Armed Forces. With benefit of hindsight, the way in which
this enormous change was implemented may positively have
encouraged litigation because of the slowness of the
adaptation at simple, administrative levels. It is not surprising
that the process was slow though, because, apart from
adaptation in institutional, not to say individual, attitudes to
fundamental change in the whole character of American
soldiering, the vast US Armed Forces' organisational
structures meant that it would take years for the changes to
leach through every area of military practice to complete,
satisfaction.

It is fair to say that something of this affected attitudes
across the Atlantic. Eventually, other NATO armed forces
followed the trail towards female integration, including those
in Britain.

From all the foregoing, therefore, it is possible to begin to
understand why women have continued to be recruited to
the British armed forces throughout successive, and
occasionally savage, personnel *cuts*, applied over and above
perceptions of male shortage. This apparently paradoxical
process seems to reinforce the idea put forward earlier that
something more powerful and quite deep is occurring with
female recruitment here. Could non-discriminatory
legislation itself be a symptom of an instinctive desire to

prepare for future, possibly large-scale conflict?

Considering how close the British came to disaster during 1939–45, can the protectiveness of the War Cabinet's policies for volunteer and conscripted women really have been justifiable, or is it an improper use of hindsight even to broach this question? What was it that made its members persist with their protectiveness for so long, and was this at men's expense?

WOMEN AND WAR, OUR ETHICAL HERITAGE

There has long been a profound belief among many men and women in this country that a special relationship exists between women and non-violence. Something of this still remains in British perceptions of the female role – some elements held in common with many religious and social philosophies brought by immigrants into these islands from different cultural and intellectual sources at various times. Chivalric and Christian ideas have exerted a potent influence.

By the early part of the twentieth century, they underlay much heart-searching about the challenge civic rights and freedoms for women presented to what was then regarded as the nature of womanhood. It had, on the other hand, been observed that other social attitudes worked against survival of mothers and children when, for example, bereft of the main breadwinner.

The experience of total war therefore consolidated the outcome of a Victorian, philanthropic pursuit of female emancipation by men of goodwill, not only by the women's reform organisations. Then the first uniformed, non-combatant women went to France in support of the British armed forces in 1917, the year after the Somme and Jutland, when the accumulated effects of so many serving in the war were driven home.

World War I, in fact, showed there might be an ethical league table, when it came to defining limits to women's new employments and the degree of danger they could carry. Where national survival could be presumed to take precedence over every other consideration, instinctive reluctance eventually made way for the practical benefit. But

British limitations to female combatancy only began to give way in 1944, again in response to pressure from manpower losses and with confidence that, finally, the threat from Germany was receding as the defensive line moved some way beyond our shores.

THE LIONESS CONCEPT

Perceptions of female aggression in relation to notions of defence are far reaching. The filter for these perceptions in modern times is created by the circumstances of women's employment in military units, for example, not only as direct combatants in territorial or civil wars but also in peacekeeping and violence-restraint as in Eritrea, Kosovo or Northern Ireland. Such perceptions carry on through to the minutiae of relationships between men and women in working military groups, involving such matters as strength differentials and the continuous puzzle as to how best to preserve an intrinsic formidability in mixed military groups.

The view through the filter takes account of the fluidity of the front-line in advanced high-tech conflicts and breaks down from there into worries over female effectiveness in hand-to-hand fighting and likely endurance capacities for the physical taking and holding of ground installations.

By comparison, for British wartime generations, non-traditional civilian work, let alone military involvement by women on a vast scale was almost as enormous to contemplate, as the notion of direct female involvement on the battlefield are to us. It required an equally deep adjustment in attitudes. For us, present-day warfare and military technology have altered so much, that we have to confront our feelings about female combatancy and how to organise reasonable prospects for survival in a highly-complex war environment.

Both then and now people are broadly comfortable with the idea of women's aggression being exerted in defence. So deeply ingrained is this belief, that many men cannot perceive the possibilities of women's adaptive capacity when they join military groups, where the intrinsic philosophy derives from notions of attack and pre-emption, rather than a defensive response.

A many-layered defensive vision nevertheless generated confidence between the sexes, reducing an incapacitating ambivalence, when, with much hard work it was successfully conveyed into the public domain by the British military authorities during the 1940s. Defence of the country against external threat was a hugely unifying force. This meant both sexes had an understanding of what was required to survive, which served to balance those conflicting attitudes and feelings just mentioned.

Today, a sense of national crisis is obviously absent, yet is there a sense of threat? Otherwise, why does the wisdom of employing women in infantry battalions keep emerging at intervals in our national conversations?[14] It seems a little too simplistic to ascribe the nagging return of this question to extreme feminist pressure, whatever that is. Could feminism itself, with all its disturbing behavioural and attitudinal manifestations, be a means by which women are subconsciously preparing themselves to manage future demands on them by society?

FEMINISM AND FEMININITY

Senior female policy-makers, thinking through the expansion of Women's Services during 1938–43, as preoccupied as everyone else was with survival, were united in their determination that women should remain feminine, even while they began to carry out male tasks and be given trousers to wear. (Remember the symbolism trousers have for sex-identification in our culture.) This was to allay public anxiety over the whole substitution scheme, but also had the effect of maintaining national morale. While women could be allowed to care about their appearance, things could not be totally bad.

Publicity photographs for the 1940s women's auxiliary services, though, did much to create an ambiguous identity for the young women. Masculine attributes in female appearance and dress tended, under the carefully-angled camera lens, to suggest competence. Photographs of Mrs, later Dame, Vera Laughton Matthews, Director of the Women's Royal Naval Service (WRNS) from 1939–46 show someone of warm and humorous personality with somewhat mannish, firm features. Nevertheless, in spite of, or perhaps because of these attributes, as well as the effect of the true depth of her character, she won the respect of many of her male peers and superiors in the Admiralty.

More difficult was the life of the Chief Controller, Auxiliary Territorial Service (ATS). Setting unsurpassed standards for turn-out and glamour, with a gift for training, supervision and public relations, Jean Knox, later Lady Swaythling, was sensitive to much within the traditional codes of behaviour applied in British military groups, for reasons that have been rehearsed in our own times. Therefore she was completely against romantic liaisons within the different mixed units which formed part of her rapidly-enlarging Service and which had attracted unwelcome criticism. She drew on cultivation of high professional standards as a means of keeping sex out of the workplace, and from influencing what went on there. Throughout her ATS units and detachments she was pretty successful in this and, with her staff, established efficiency and order out of the initial organisational muddle afflicting all the Services, not only the women's. She herself wished for no other basis for judgment of her own abilities than good results.[15]

Her predecessor, Dame Helen Gwynne-Vaughan, of formidable intelligence, scholarly and administrative ability, who had taken that first Women's Army Auxiliary Corps (WAAC) contingent to France, famously described in her memoir a visit to the War Office on 13 February 1917 to be interviewed for the task by the Adjutant General: 'I was excited and alarmed, but I still remember the very excellent little tricorne [hat] of black panne which graced the occasion and helped to give me confidence ...'[16]

Interviewing WRNS air mechanics at RNAS Culdrose,[17] this writer was struck by some of the young women's boyish mannerisms, and with their extremely radical haircuts. Now, over 20 years on, their style can be seen everywhere among women of all ages. The young Wrens still looked female, but in the detail of some of their behaviour, (e.g. movement, speech, diction, directness of glance) some signs of masculinity were noticeable. Undoubtedly these mannerisms enabled more efficient communication with the men. They related very firmly in a largely sisterly fashion to their male colleagues.

On the other hand, among the early sea-going WRNS air mechanics highly feminine, latent tribal instincts were also detectable within their own descriptions of their experiences. They learned to stick to 'their' Royal Naval males, expressing great wariness of the unfamiliar men of the Royal Fleet Auxiliary with whom they sailed. The RFA seamen seemed threatened by female presence and had little appreciation of the carefully-constructed working relationships of the women and men of the Royal Naval group.

Even so, with such wide interest in the armed services in this country, whatever the ups and downs of public regard in which they are collectively periodically held, would something along the lines of the reactions of the RFA seamen explain why it taken so long for British women to be officially commemorated for their commitment to the national interest during World Wars I and II?[18] Women everywhere were proportionately less decorated for valour than men.

After 1945 returning Soviet women veterans kept quiet about their experiences as their poor reputation for roughness and immorality affected chances of marriage.

Did those women, by taking on men's roles in war, jeopardise their own sexual identities within their communities?

Did they compromise their apparent potential as future mothers?

Could it be that, against respect for the female war contribution, there exists an uneasy recognition by both sexes that the traditionally-male effort was simply not enough?

Perhaps there is more than Norman Dixon's theory of the emasculation of male-specific tasks by female adoption[19] here. Is there a relationship between female presence in hitherto all-male military groups and a sense of threat to our future viability as a people which produces this comprehensive perceptual shut-down?

By 1945, when British men were demobilised from military service in their thousands, women returned from the workplace into the home to make room for them. By way of contrast a Chinese proverb, one among many similar reflecting centuries of female subordination, emerged during the 'Great Leap Forward' of the 1950s as women entered the workforce on a large scale: 'A woman having a job is like flying a kite under a bed.'[20] Chinese women's difficulties at that time have been ascribed, however, to slow adaptation to new social demands by the women themselves, as well as men.

MANPOWER – READING THE RUNES

A profound sense of manpower shortage, therefore, is an abiding impression left after this writer's immersion in the British War Cabinet papers and in the military and civil archives of World War II. Given the general sense of pressure on numbers, why did the War Cabinet bother to go to such lengths devising ways and means for survival which effectively restricted an all-out manpower supply? It seems likely that this was because several of its members and its political, military and scientific advisers, who had lived through or seen action in World War I, were imbued with the idea that survival was possible if involvement in the carnage was limited. The overall effect on the prospects of the fighting male troops of the restrictions or exemptions surrounding manpower distribution is hard to gauge. It could be reasonably implied that more troops in theatres overseas were able to rest as the result of female substitution at home however restricted, in spite of some notorious cases of overuse and exhaustion.

Embodied perhaps by Churchill, the War Cabinet's deliberations on every aspect of the management of the war, therefore, exude a determination to survive, to remember normal life and to reintroduce its mechanics as soon as possible after the cessation of hostilities.[21] Its members resisted the profound psychological pressure of dependence on numbers and only altered their rules of involvement when they considered they had no alternative. This might also explain why some units suffered from over-use: and why some of the manpower management detail will not ever make sense except in this context of measured relaxation of restrictions.

Gradually, however, some restrictions were relaxed as manpower pressures grew heavier. Married women could volunteer if they were able to show their dependents and children were being cared for. Young and older men, hitherto excluded, were employed in combatant jobs. Female gunners were asked to go as volunteers to the Continent with mixed

AA regiments in 1944, while some Wren codes and cipher specialists served in convoys at sea.

DIFFERENCES OF OPINION — SOLVED BY OBSERVATION

There was a difference of opinion, in terms of ethical appreciation, of what was needed to survive – among administrative and governing classes, as well as elsewhere in the British social spectrum during 1939–45. Changes of mind about the employment of women in uniform only occurred among those who saw reason for it and came to understand how valuable a female contribution to the war effort was.

That reluctance to come to terms with the all-encompassing effects of the war and admit our closeness to disaster, by accepting the fact of having actually to conscript women as military auxiliaries, continued well into the 1960s. This invites the conclusion that among those men who found acceptance difficult, feelings must have been rooted at a level beyond the rational. No-one could possibly ignore the very great success of the, then, WRNS, ATS and WAAF in delivering all that was asked of them, in spite of the considerable range of serious social and practical problems they had to contend with.

The mechanisms of the wartime government were designed, and evolved, to take fair account of public feeling and attitudes. These provided some checks and balances which, although incredibly stressful for those involved in working them through, seem by a rough rule of thumb to have been capable of adaptation in the teeth of every kind of opposition when the pressure was heavy enough. From the wartime archives, it is clear that a struggle went on which reflected this within the British civil and military war administration, virtually at every level. The struggle was exemplified in the pressures the Treasury was subjected to in the fierce competition for scarce or non-existent resources.

Some Treasury officials come across the years from their wartime memoranda or minutes as deeply unpleasant people – an impression which must be corrected by the recognition that they evidently felt extremely beleaguered. The tone of many letters emanating from Treasury sources is frequently aggressively-defensive. General Sir Frederick Pile's memoir, enlightening about much, is particularly instructive about this. It is also extremely instructive about the whole ethical question of the employment of women in combatant or semi-combatant jobs, the stance that he took, and why – much of what is discussed or described in his book still has strong resonance for us today.[22, 23]

A veteran of World War I by which, along with most of his generation, he was deeply affected,[24] General Pile was among the earliest to conceive the idea of employing women with anti-aircraft guns. In 1938 he consulted a female engineer, Caroline Haslett, about its feasibility. He was thoughtful from the start about the women's health and safety. He bore in mind throughout the war that the female auxiliaries he was employing in AA Command were linked to families who required to be reassured, and the maintenance of whose morale was essential for national cohesion.

CODA

The War Cabinet's members had no hesitation over limiting the combatant involvement, not only of older, or younger women with dependents, but also of groups of men. They exercised careful husbandry in conjunction with the Ministry of Labour, of the different age-groups of both sexes for their direction to industry, agriculture, home defence or the armed services. Their judgement was ultimately tested by their choice of timing.

Through the expression of contradictory attitudes for and against the employment of women throughout the wartime infrastructure, the British public signified a deep reluctance to become totally militarised. Although slowness to adapt undoubtedly cost many male lives as the struggle for organisational efficiency went on into 1942, the perpetual tension between profound conservatism and rapid adaptation to the exigencies of war nevertheless resulted in a certain crude balance between demand and supply. Even so, early adaptation to the war's requirements was slow enough to make it a close-run thing.

Unlike our predecessors of 1938–45, though, we probably will not be able to work out such deep differences of opinion 'on the hoof' during engagement in any future serious conflict. There simply may not be the time.

While considering what the substance of our survival in a nuclear, biological and chemical age amounts to, we could perhaps do worse than identify further ways of delivering reassurance to military men and women alike, following the example of gender-free testing, and showing the virtue of developing the relative strengths and skills of each individual. Flexible adaptation and capitalisation on aptitudes has long been practised by our modern armed services in training women, as well as men – indeed the biggest argument for avoiding sexual stereotyping is the retention of the maximum possible flexibility.

The young WRNS air mechanics working at Culdrose 20 years ago on whom the pilots depended, wore punk haircuts, trousers and, after requests, specialist tough footwear suitable for wet decks. They were fairly small and compactly-built, being selected for aptitude before height – fit, bright and alert. They saw no contradiction between their work and their femininity. They were instinctively resourceful in their approach to their work problems and with their male comrades made appropriate and flexible use of each other's skills. At the end of their shifts the young Wrens went home to fall into a deep sleep before going out on the town with their boyfriends. Devoted to their task, they were nevertheless unromantic realists.

The inner sparkle they displayed, familiar to all armed services' recruiters, and also found by this writer again among army recruits in Pirbright not long ago, was described by Helen Gwynne-Vaughan in the final chapter of her memoir nearly 60 years before. She detected in it an energy and enterprise, in the young women she had encountered since 1917, which came with a readiness to face hard conditions and to do impossible things. She saw it as a priceless national asset, not to be treated lightly.

Infantry training to certain levels is necessary, let it be said to any non-military reader, for everyone working in the combatant specialities. The variety of field skills it provides makes trained individuals more, not less, effective operators at many levels. In any case, some infantry skills come automatically as part of the various training packages.

Therefore, it would plainly be wrong to regard infantry training in all but its higher reaches as an isolated event, with which the rest of the army's branches or specialties are unfamiliar.

Interchangeability of roles is always happening in one way or another in war and is frequently related to personnel shortage – lack of reinforcements. Thus flexibility can, and has already been found to be, of immense value – without it events during a hostile encounter can occasionally become 'somewhat thought-provoking'.[25]

This is where the introduction of women to infantry training becomes important for their effectiveness by adding to their skills, as much as this does for men – with life-preserving as much as life-taking potential for both.

Most of the questions asked within this essay have been answered directly or by implication. If sex differentials in strength between men and women are so different as to cause the men genuine anxiety, then all-male platoons should remain for reassurance of both sexes. Infantry skills and their application across the whole spectrum of modern warfare must not, however, be denied to servicewomen, in the interests of developing their training to its fullest potential for the benefit of male colleagues, as well.

The fundamental ethical query has also already been answered – women are still frequently required to make good male recruiting shortfall in the short-term so as to prepare reasonably for a future crisis. National interest, however, ultimately will prevail over some ethical values in a complex interaction. Training military women to take life in peacetime provides a core of knowledge, however hard, of the realities of war and a means of standing up to its pressures. We have seen how it is that if survival itself becomes the greatest pressure and numbers critical, then ethical attitudes will always tend towards the pragmatic, as they eventually did during 1939–45. In future, they may do so more swiftly. The apparent blurring of male and female roles (we would say feminisation or masculinisation) throughout our society during that period has steadily reasserted itself, after a pause between 1950 and 1970.

It has been suggested, and has emerged in conversations between this author and veterans from World War II, that the idea of defending hearth and home is a potent stabiliser in difficult times; and that the presence of women is an intrusion into definition of maleness. While true for some, it is also likely that for other men, observing how genuine comradeship between the sexes was built up within the ADGB organisation during the 1940s, that male edginess in this respect has other roots. Perhaps it stems from perceptions of a requirement for a post-war refreshment, for rest of mind and body; from a search for a route back to stability; from a need for the decency if not the innocence of family life, or for the cheerfulness of returning to old pursuits – which a female infantry presence would interrupt.

Much has been made of atrocities carried out both during and after warfare, and post combat cruelties of every kind constitute a form of ancient male behaviour which cannot be pushed aside. However, in well-disciplined troops, with a clear sense of purpose, who maintain their organisational structures into the aftermath of combat, there is less likelihood of anarchy, though its possibility even here cannot be excluded and must be prepared for. The Russian atrocities following their victory in Germany were well known – and it is clear that this is because control of the troops was lost at a critical time. Traditionally, men are good at compartmentalising their activities at practical and emotional levels – an intrinsic capability for war may form the foundation for this skill.

Modern studies in biological determinism have done much to illuminate aspects of male-female behaviour and if some ideas can be applied in a socio-military context, it is, then, possible that some solutions to controlling post-combat mayhem might be discoverable, also. Further, post-war rewards would have to be as clearly delineated as the capture of enemy women has traditionally been. (Women in such situations are not, however, generally assumed to be armed.)

If women are traditionally excluded from the 'war' compartment of the male mind as a means to men's own survival; if our collective social perception of war and warfare hitherto has been male-centric for historical reasons, then with the persistence of recruitment of women to our armed services, then our developing understanding of the nature of future war must be seen in the light of the lethality of modern small arms, or extensive or intricate weapons-systems. This perhaps illuminates our understanding of the psycho-social origins of the leading edge, being formed by American approaches to modern war, which aim for reducing rather than increasing numbers-involvement. One senses another paradox lying in wait for us.

The whole question of male-female/female-male role perceptions therefore is deep and subtle and the Royal Navy and RAF are, as the result of long practice, well ahead of the British Army in this area. Their route to mixed-sex working and fighting (I apologise for this pun) has been easier because it cannot be argued that naval women or airwomen and women officers are inevitably going to come into direct, physical combat with an enemy. Women and men in both these Services are in analogous situations. It really can be argued that, for them, direct contact is less rather than more likely, whereas for ground forces, even for soldiers employed in the logistic chain, the probability of a face-to-face hostile encounter is universally accepted as being far higher. Distance, therefore, between belligerents is vital to acceptance, by some, of combatancy for women in whatever form it may take; a profound need for a defensive effort is another.

Apart from recruiting shortages, another reason why the Royal Navy has stuck by its policy of training women to serve at sea and, indeed to award women sea commands (first in the reserve and then in the main organisation), and why the RAF has remained faithful to its recruitment of female fighter and other pilots, is because the men of both Services have had long experience of putting their lives into female hands. Mutual trust, therefore, has been long established on practical experience.

In the Army, however, the ethos has been quite different. Since 1945 there has been no comprehensive tradition of women sharing either hardship or danger with men. Considerable extension of jobs for women in the Army since 1970, and again since the early 1980s, has been slowly taking place against an increasingly complicated background of legislative demand; high personnel wastage, caused by cuts and voluntary redundancies, and sharp and inconvenient fluctuations in recruiting.

Now that a little more defence funding is on its way, and a decent period for reinforcing consolidation of job extensions, new ideas in the distribution of employments between the sexes might now be possible. It is clear from this country's broadest experience of war since 1917, that servicemen and women have, can, and do work excellently together and can provide much mutual, professional reassurance and support, achieving levels of high efficiency, under stressful and highly-threatening conditions.

That women in uniform may also have a capacity for gallantry, as demonstrated by some of our own wartime auxiliaries and, more recently, by notable American examples, is now becoming increasingly accepted in our own society and is another aspect of our complex and difficult legacy from an era of total war.

REFERENCES

This chapter is an expanded version of an entry for the Trench Gascoigne Prize Essay Competition for 2000 in which it took 3rd place. Thanks are due to the Royal United Services Institute for Defence Studies for its release from copyright.

1. House of Commons Written Answers for 22 May 2002 (pt 11) Column: 364W *Service Personnel,* Secretary of State for Defence, <www.parliament.the-stationery-office.co.uk>; Ministry of Defence, Directorate of Service Personnel Policy Service Conditions *Women in the Armed Forces,* Summary of Report by Steering Group, London, May 2002, In Ministry of Defence, *Women in the Armed Forces* Report by Employment of Women in the Armed Forces Steering Group; 15pp supporting Annexes A-E, London, May 2002.
2. Margaret Phillips, *Small Social Groups in England* (London: Methuen 1965).
3. Beryl E. Escott, *Women in Airforce Blue* (Wellingborough: Patrick Stephens 1989) pp.104, 160–4; Eric Taylor, *Women Who Went to War* (London: Robert Hale 1988) (illus. set within pp.208–9) showing ATS member armed, 'somewhere in NW Europe', evidently with Sten Mk3. Anecdotal evidence regarding lack of [official] weapons-training for ATS in the five mixed AA regiments sent to secure Antwerp and Brussels 1944, comes from author's correspondence March–April 1983 with Major-General B.P. Hughes, who was BGS (Operations) to Gen. Pile, AA Command, 1942–44. The General kindly consulted his sister who had served there, and stated no training of ATS in use of firearms was ever carried out. [This affronts a basic military ethic of never sending a soldier unarmed into a dangerous situation.] General Hughes' comment: 'How you square this with the fact that they operated a predictor, the sole purpose of which was to direct something nasty at the enemy, I don't know!'
4. See Georgina Natzio, 'British Army Servicemen and Women 1939–45: Their Selection, Care and Management', *RUSI Journal* (London Vol.138, No.1, (February 1993) pp.36–43: (some refs misprinted, corrections available from author via RUSI).

5. *WAAF substitution in the Balloon Command*, Report by Air Marshal E.L. Gossage on successful experiment in training: 100% female balloon crews, 30 September 1942, AIR 2/4710 PRO, London.
6. Squadron Leader E.A. Natzio, RAFVR, Conversation with author, 1983.
7. Anne Eliot Griesse and Richard Stites, 'Russia: Revolution and War', in Nancy Loring Goldman (ed.) *Female Soldiers – Combatants or Non-combatants? Historical and Contemporary Perspectives* (Westport, CT and London: Greenwood Press 1982) pp.61–85.
8. Antony Beevor, *Stalingrad* (Harmondsworth, UK: Penguin 1998).
9. Richard Overy, *Russia's War* (London: Allen Lane the Penguin Press 1998) pp.241–2, 287–8, 359 n44, 373.
10. Ibid.
11. J.M. Cowper for the War Office: *Auxiliary Territorial Service* (London: HMSO 1949) p.iii: Appendix VI *Analysis of strengths on 30th September 1943, by Trades and Employments and by Arms, etc with which employed* , WO 277/ 6 PRO London.
12. Sun Tzu, *The Art of War*, translated Samuel B. Griffith (New York and Oxford: Oxford UP 1963) pp.7, 72–5. Does this famous remark, plus footnote, suggest a preoccupation with numerical superiority? By contrast, in another translation, by Ralph D. Sawyer (Boulder, CO: Westview Press 1996) p.173, the same passage is translated, and laid out on page in a manner that seems to highlight the cost and sustenance of large operations, but perhaps this is fanciful, as both matters are dealt with in each translation, if textually slightly differently
13. Martin Binkin and Shirley J. Bach, *Women in the Military* (Washington DC: Brookings 1977) pp.13–14.
14. Nigel Foster, 'Demand for women in regiment unmonstrous', *The Independent* (London), 21 November 1998.
15. Conversations and correspondence, Lady Swaythling, formerly Jean Knox, Chief Controller, ATS 1941–43, during 1982–1990.
16. Dame Helen Gwynne-Vaughan, *Service with the Army* (London: Hutchinson 1942) p.14.
17. Interviews, WRNS air mechanics, RNAS Culdrose, Cornwall, 26 July 1982.
18. Press reports: 'Women at war favourites for Square Statue', *The Independent*; 'Women in War, plan for plinth', *The Times*; Mick Hume, 'The battle of the Trafalgar Square statues reveals ours as a small-minded age in which heroes and heroism are out of fashion', *The Times*; all 14 February 2000.
19. Norman Dixon, *On the Psychology of Military Incompetence* (London: Jonathan Cape 1976) pp.208–13.
20. Elisabeth Croll, *Women in Rural Development: The People's Republic of China* (Geneva: International Labour Office 1979) p.37.
21. War Cabinet Minutes and Conclusions Series London: PRO, 1938–45.
22. Treasury Files, London PRO 1939–45. cf. War Office files and War Cabinet papers.
23. General Sir Frederick Pile, *Ack-Ack – Britain's Defence against Air Attack during the Second World War* (London: Harrap 1949) p.24.
24. Colonel Sir Frederick Pile (son of above), Conversation with author, 1983.
25. Colonel, later General Nick Vaux RM, Conversation with author, 1985.

Index